Probability Theory

Probability Theory

Proceedings of the 1989 Singapore Probability Conference held at the National University of Singapore, June 8–16, 1989

Editors

Louis H.Y. Chen
Kwok Pui Choi
Kaiyuan Hu
Jiann-Hua Lou

Walter de Gruyter · Berlin · New York 1992

Editors

Louis H. Y. Chen
Kwok Pui Choi
Kaiyuan Hu
Jiann-Hua Lou
Department of Mathematics, National University of Singapore,
Lower Kent Ridge Road, Singapore 0511, Republic of Singapore

1991 Mathematics Subject Classification: Primary: 60-06; 58-06. Secondary: 05C40, 05C80; 35J10; 58G32; 60F05, 60F10, 60F99, 60G15, 60G60, 60G99; 60H07, 60H15, 60H25, 60H30, 60H99, 60J35, 60J65; 62M40.

♾ Printed on acid free paper which falls within the guidelines of the ANSI to ensure permanence and durability.

Library of Congress Cataloging-in-Publication Data

Singapore Probability Conference (1989 : National University of Singapore)
Probability theory : proceedings of the 1989 Singapore probability Conference held at the National University of Singapore, June 8-16. 1989 / editors, Louis H. Y. Chen ... [et al.].
p. cm.
Includes bibliographical references.
ISBN 3-11-012233-2
1. Probabilities--Congresses. I. Chen, Louis H. Y. (Louis Hsioa Yun), 1940- .
QA273.A1S6 1989 92-14169
519.2--dc20 CIP

Die Deutsche Bibliothek - Cataloging-in-Publication Data

Probability Theory : proceedings of the Singapore Probability Conference held at the National University of Singapore, June 8-16, 1989 / Ed. Louis H.Y. Chen ... - Berlin ; New York : de Gruyter, 1992
ISBN 3-11-012233-2
NE: Chen, Louis H. Y. [Hrsg.] ; Probability Conference <1989, Singapore> ; <National University, Singapore>

Printed in Germany.
Printing: Gerike GmbH, Berlin. Binding: Lüderitz & Bauer GmbH, Berlin.
Cover design: Thomas Bonnie, Hamburg.

Preface

This volume contains most of the workshop and invited lectures given at the Singapore Probability Conference in the National University of Singapore from 8 to 16 June 1989. The conference was jointly organized by the Singapore Mathematical Society and the Department of Mathematics of the National University of Singapore. It was sponsored by the Southeast Asian Mathematical Society and held under the auspices of the Singapore National Academy of Science.

We are grateful to the workshop and invited speakers for their stimulating lectures; in particular, to Professor Kiyosi Itô for delivering a Special Invited Lecture. Our thanks also go to the speakers of contributed talks for their active participation and to all the participants for making the conference an unforgettable event. We are particularly thankful to Professors Takeyuki Hida, Harry Kesten, David O. Siegmund and Marc Yor for their invaluable advice during the planning of the conference and the preparation of these proceedings. We would also like to acknowledge with appreciation the generous financial support from Mr. Ng Kong Beng, Lee Foundation, UNESCO, Shaw Foundation, Tan Kah Kee Foundation, Hong Leong Foundation and Shell Eastern Petroleum (Pte) Ltd.

We are thankful to the National University of Singapore for the use of its facilities and to the non-academic staff of the Department of Mathematics for their excellent administrative support. We would also like to place on record our appreciation to the staff of Walter de Gruyter & Co. for their patience, understanding and cooperation. Finally we would like to thank friends, colleagues and all those who have helped in one way or another in making the conference possible.

Louis H. Y. Chen
Kwok Pui Choi
Kaiyuan Hu
Jiann-Hua Lou

Advisors

Takeyuki HIDA

Harry KESTEN

David O. SIEGMUND

Marc YOR

Organising Committee

Chairman	Louis H. Y. CHEN
Secretary	Jiann-Hua LOU
Committee Members	Yiu Man CHAN
	Kwok Pui CHOI
	Kaiyuan HU
	Fred Pui Fai LEUNG

Table of Contents

Appendices

The following lectures are not included in this volume:

Workshop lectures

P. DIACONIS
Group representations, stopping times, and eigenvalues of the Laplacian

E. B. DYNKIN
Branching and superprocesses

Invited lectures

J. AZÉMA
A new class of martingales with chaotic representation property

A. D. BARBOUR
On Stein's method for diffusion approximations

P. DIACONIS
Geometry and Markov Chains

E. B. DYNKIN
Superprocesses and partial differential equations

C. R. HWANG
Simulated annealing

D. W. STROOCK
Spectral gap and logarithmic Sobolev inequalities

SINGAPORE PROBABILITY CONFERENCE
WORKSHOP : 8 - 10 JUNE 1989 LT 21
CONFERENCE : 12 - 16 JUNE 1989 LT 21. 20
ORGANISED BY DEPARTMENT OF MATHEMATICS, NUS
SINGAPORE MATHEMATICAL SOCIETY

1. CHIEW Mui Kui
2. Margaret STEIN
3. EE Soo Mei
4. LIM Siok Wan
5. KHATIJAH Syed Ahmad
6. CHOU Ching-Sung
7. M. Gopalan NAIR
8. LING Kiong Doong
9. Adolfo M. DE GUZMAN
10. HU Kaiyuan
11. HE Shengwu
12. CHENG Kuang-fu
13. Jose S. VICTORIA
14. YEO Kwee Poo
15. Susanti LINUWIH
16. Kiyosi ITÔ
17. Louis H. Y. CHEN
18. LIN Ta-feng
19. Karl SIGMAN
20. XIE Shengrong
21. GONG Guanglu
22. LIU Xiufang
23. CHIANG Tsuu-Shuh
24. POH Kheng Siong, Roger
25. Shinso WATANABE
26. Jong Woo JEON
27. Kun Soo CHANG
28. Persi DIACONIS
29. Mark A. PINSKY
30. C. C. HEYDE
31. Timothy BROWN
32. A. D. BARBOUR
33. Lars HOLST
34. Malcolm QUINE
35. Andy H. LEE
36. N. N. CHAN
37. SHEN Anqi
38. LAM Yeh
39. Jiann-Hua LOU
40. DAI Yonglong
41. HOU Zhen-ting
42. Harry KESTEN
43. LIU Wen
44. Charles STEIN
45. E. B. DYNKIN
46. Daniel W. STROOCK
47. Angela WADE
48. Rm. SEKKAPPAN
49. S. R. ADKE
50. CHAO Min-Te
51. Takeyuki HIDA
52. Dong Myung CHUNG
53. J. AZÉMA
54. Shinichi KOTANI
55. W. J. ANDERSON
56. CHANG Boon Chong
57. SHEE Sze Chin
58. HUANG Zhirui
59. CHAN Yiu Man
60. GAN Fah Fatt
61. LEUNG Pui Fai, Fred
62. NG Tze Beng
63. CHAN Onn
64. TAY Kian Boon
65. LOH Wing Wah
66. MAK Kok Sun
67. MA Siu Lun
68. TANG Wai Shing
69. LEUNG Ka Hin
70. Roger TAN
71. PANG Yu Hin, Peter
72. LO Shaw-Hwa
73. Shisu NAKANISHI
74. YANG Zhen Ming
75. Kritsana NEAMMANEE
76. Yasushi ISHIKAWA
77. Hiroshi ISHITANI
78. David O. SIEGMUND
79. Shigeo TAKENAKA
80. YAN Jia-An
81. XU Jia-Gu
82. Minoru MOTOO
83. Mamoru KANDA
84. K. Joseph MOURAD
85. CHOI Kwok Pui
86. Narn-Rueih SHIEH
87. Robert C. DALANG
88. HWANG Chii-Ruey
89. Paolo BALDI
90. CHOW Yunshyong
91. Masami OKADA
92. WONG Woon Kwong
93. LEE Soo Teck

Workshop lectures

Inverse questions in stochastic differential geometry

Mark A. Pinsky

Probability Theory

Lecture I : Brownian motion and Riemannian geometry

I.1. Geometric background

We work in the context of *complete Riemannian manifolds*. This means that we are given two objects : (i) a differentiable manifold M of dimension d and (ii) a Riemannian metric $<,>$. The latter operation produces a smooth function $m \to < X, Y >_m$ for each pair of smooth vector fields (X, Y) on M. At each point $m \in M$ the operation $(X,Y) \to < X, Y >_m$ defines a Euclidean inner product on the tangent space M_m, so that in a suitable system of coordinates near m we may write $< X, Y >_m = \sum_{i=1}^{d} x_i y_i$. The *unit tangent bundle* $SM := \cup_{m \in M} S_m$ has a natural manifold structure, where S_m is the unit tangent sphere $S_m := \{\xi \in M_m :< \xi, \xi >_m = 1\}$; locally this may be viewed as a product $M \times S^{d-1}$, which we will often do. There is a *canonical projection operator* $\Pi : C^\infty(SM) \to C^\infty(M)$ defined as follows: The unit tangent sphere has a unique measure ω_m which is rotationally invariant with respect to the orthogonal group $O(S_m)$ and is normalized so that $\int_{S_m} 1 \omega_m(d\xi) = 1$, which may be called *normalized Lebesgue measure.*

The canonical projection operator $\Pi : C^\infty(SM) \to C^\infty(M)$ is defined by

$$\Pi f(m) = \int_{S_m} f(m, \xi) \omega_m(d\xi).$$

It satisfies the properties that $\Pi 1 = 1$ and $\Pi^2 = \Pi$; the range of Π is $C^\infty(M)$.

A *Riemannian connection* ∇ is an operation on pairs of vector fields $(X, Y) \to \nabla_X Y$ which is additive in each component and satisfies the differentiation rules

$$\nabla_{fX}(gY) = f(Xg)Y + fg(\nabla_X Y)$$
$$X < Y, Z > = < Y, \nabla_X Z > + < \nabla_X Y, Z >$$

for each pair of smooth functions f, g and triplet of smooth vector fields X, Y, Z. The fundamental lemma of Riemannian geometry asserts the existence of a Riemannian connection which is unique under an additional condition of symmetry. A smooth curve $t \to \gamma(t) \in M$ is a *geodesic* if we have $\nabla_{\gamma'(t)} \gamma'(t) = 0$. It is proved that this notion is well-defined and that geodesics exist locally, i.e. for each $m \in M, \xi \in M_m$ there exists a geodesic $\{\gamma(t), -\delta < t < \delta\}$ such that $\gamma(0) = m$, $\gamma'(0) = \xi$. The length $< \gamma'(t), \gamma'(t) >^{\frac{1}{2}}$ is constant along any geodesic. The Riemannian manifold is said to be *complete* if we can choose $\delta = +\infty$, i.e. geodesics are globally defined.

The *canonical horizontal vector field* D on SM is defined by

$$Df(m, \xi) = (d/dt) f(\gamma(t), \gamma'(t))|_{t=0}$$

where γ is the geodesic with $\gamma(0) = m$, $\gamma'(0) = \xi$. [In a system of the local coordinates $(x_1, ..., \xi_d)$ this may be computed by the formula $D = \xi_i \partial/\partial x_i - \Gamma^k_{ij}(m) \xi_i \xi_j \partial/\partial \xi_k$, where we sum on repeated indices.] (Γ^k_{ij} are the "connection coefficients", given by $\Gamma^k_{ij}(m) = < \nabla_{e_i} e_j, e_k >_m$). This is a globally defined non-zero vector field on SM which is related to the canonical projection operator Π by the relation

$$\Pi D \Pi = 0$$

reflecting the fact that any linear function on S_m has integral zero.

In the second lecture we will need the *exponential mapping* which we now define : $exp_m : M_m \to M$ is the smooth mapping which sends a vector $\xi \in M_m$ to $\gamma(1)$, the geodesic with $\gamma(0) = m$, $\gamma'(0) = \xi$. We have the identity $\gamma(t) = exp_m(t\xi)$, which may be used to show that for small t we have the isometric property that dist $(\gamma(t), m) = t < \xi, \xi >^{\frac{1}{2}}$.

I.2. Isotropic transport process

Having defined the necessary quantities from geometry, we now construct a canonical Markov process on the unit tangent bundle.

Let $\{e_n\}_{n \geq 1}$ be a sequence of independent and identically distributed random variables on a probability space (Ω, F, P) with the distribution

$$P\{e_n > t\} = e^{-t} \quad (n \geq 1,\ t > 0).$$

The partial sums $t_n = e_1 + ... + e_n$ are the jump times of a Poisson process with unit rate.

For a given starting point $(m, \xi) \in SM$ let

$$(X(t), \xi(t)) = (\gamma(t), \gamma'(t)) \quad (0 \leq t < t_1).$$

The random variable ξ_1 is required to be conditionally distributed according to normalized Lebesgue measure, i.e.

$$E\{f(\xi_1)|e_1\} = (\Pi f)(X(e_1)).$$

Assuming that $\xi_1, ..., \xi_{n-1}$ have been defined, we require that

$$E\{f(\xi_n)|e_1, ..., \xi_{n-1}\} = (\Pi f)(X(t_n))$$

In the time interval $t_n \leq t < t_{n+1}$ we define

$$(X(t), \xi(t)) = (\gamma(t - t_n), \gamma'(t - t_n)) \quad (t_n \leq t < t_{n+1})$$

where γ is the geodesic with $\gamma(0) = X(t_n)$, $\gamma'(0) = \xi_n$.

In this way we obtain a globally defined Markov process $Y(t) = (X(t),\ \xi(t))$ taking values in the unit sphere bundle, called the *isotropic transport process*. The sample paths are piecewise geodesic, with jumps in the tangent vector at the jump times of a Poisson process with rate one. We now compute the infinitesimal generator

$$Lf(m, \xi) := \lim_{t \downarrow 0} t^{-1}[Ef(X(t), \xi(t)) - f(m, \xi)] \quad (f \in C^\infty(SM)).$$

With probability e^{-t} we have $t < e_1$ and

$$f(X(t), \xi(t)) = f(\gamma(t), \gamma'(t)).$$

With probability te^{-t} we have $e_1 \leq t < e_1 + e_2$ and

$$f(X(t), \xi(t)) = f(m, \xi_1) + o(1).$$

All other possibilities have probability $o(t), t \downarrow 0$ and so

$$\begin{aligned} Lf(m,\xi) &= \lim_{t\downarrow 0}[(e^{-t}/t)[f(\gamma(t),\gamma'(t)) - f(m,\xi)] + e^{-t}Ef(m,\xi_1) \\ &\qquad - t^{-1}(1-e^{-t})f] \\ &= Df + \Pi f - f. \end{aligned}$$

The infinitesimal generator of the isotropic transport process is an integro-differential operator of a very simple structure on the unit sphere bundle. We simply "differentiate in the horizontal direction and integrate over each vertical fiber".

I.3 Central limit theorem

To establish the connection with Brownian notion, we introduce a small parameter $\delta > 0$ and consider a sequence of generators

$$L_\delta = D/\delta + (\Pi - 1)/\delta^2 = (1/\delta^2)[\delta D + (\Pi - I)].$$

This generates a process on the δ-sphere-bundle, time-changed by the deterministic factor $1/\delta^2$. We will show that this sequence of processes converges weakly to a diffusion process.

For this purpose we verify the following identity

$$(+) \qquad L_\delta(f + \delta Df + \delta^2 D^2 f) = \Pi D^2 f + \delta D^3 f$$

where $f(= \Pi f)$ is a given element of $C^\infty(M)$. We recognize ΠD^2 as the *normalized Laplacian* (where we sum over repeated indices)

$$\begin{aligned} \Pi D^2 f &= \Pi(\xi_i \partial/\partial x_i - \Gamma^k_{ij}\xi_i\xi_j\partial/\partial\xi_k)(\xi_\alpha \partial f/\partial x_\alpha) \\ &= \Pi(\xi_\alpha\xi_i)\partial^2 f/\partial x_i\partial x_\alpha - \Pi(\xi_i\xi_k)\Gamma^k_{ij}\partial f/\partial x_k \\ &= (1/d)g_{ij}[\partial^2 f/\partial x_i\partial x_j - \Gamma^k_{ij}\partial f/\partial x_k]. \end{aligned}$$

The computation of the Laplacian is especially simple in the case of a "radial function", meaning that $f(x) = F(r(x))$ where $r(x) = dist(x, m)$ and m is a fixed reference point of M. In this case we have the formula

$$d\Pi D^2 f(x) = F''(r) + [\Theta'(x)/\Theta(x)]F'(r).$$

Here $\Theta(x)$ is the "change of volume factor" induced by the exponential mapping : $\Theta(x) = |dexp_m(x)/dx|$ [24].

The operator identity (+) asserts that L_δ, the generator of the scaled isotropic transport process, is an approximation to the normalized Laplacian in the extended sense that for each $f \in C^\infty(M)$, there exists a sequence $f_\delta \in C^\infty(SM)$ such that $f_\delta \to f$ and $L_\delta f_\delta \to \Pi D^2 f$ when $\delta \downarrow 0$. This allows one to prove the pointwise convergence of the resolvent operators : for each $f \in C^\infty(M)$

$$\lim_{\delta \downarrow 0} R_\lambda^{(\delta)} f = R_\lambda f$$

where $R_\lambda^{(\delta)} = \int_0^\infty e^{-\lambda t} e^{tL_\delta} dt$, $R_\lambda = \int_0^\infty e^{-\lambda t} e^{t\Pi D^2} dt$ in the sense of Hille-Yosida. To make the details work one needs to know that the limiting semigroup exists and has the Feller property of preserving the class $C_o(M)$. If M is compact this is automatic; otherwise one may assume that the volume growth is "sub-exponential" $vol(B_m(r)) \le Ce^{kr}$ for positive c, k. From this one can prove the convergence of finite-dimensional distributions and eventually the weak convergence in the space of continuous functions.

The existence of Brownian notion can be established directly without recourse to an approximation method [1]. For this paper we introduce the *bundle of orthonormal frames*

$$O(M) = \{(m, e_1, \ldots, e_d) : m \in M, (e_1, \ldots, e_d) \text{ is an } ON \text{ basis of } M_m\}.$$

The *canonical horizontal vector fields* are defined by

$$L_i f(m, e_1, \ldots, e_d) = \frac{d}{dt} f(\gamma(t), e_1^{(i)}(t), \ldots, e_d^{(i)}(t))|_{t=0}$$

where $e_j^{(i)}(t)$ is the parallel transport of the vector e_j along the geodesic $\gamma(t)$ with $\gamma(0) = m, \gamma'(0) = e_i$, in particular $\gamma'(t) = e_i^{(i)}(t)$. The Brownian motion is defined by solving the Stratonovich SDE

$$dX_t = \sum_{i=1}^{d} L_i(X) o dW^i$$

where $(W^1, \ldots, W^d)$ is a standard Wiener process on $\mathbb{R}^d$. This defines a Markov diffusion process on $O(M)$ with infinitesimal generator

$$\Delta_{O(M)} = \frac{1}{2} \sum_{i=1}^{d} L_i^2$$

the so-called "horizontal Laplacian". In particular for $f \in C^\infty(M)$ we have $\Delta_{O(M)} f = (d/2)\Pi D^2 f$ [3].

I.4. Application to recurrence

The methods used to prove the central limit theorem may also be used to study the transience and recurrence of the isotropic transport process on a non-compact manifold. Recall the following Lyapunov criteria for recurrence and transience of Markov processes:

- The process is transient if there exists a positive function f defined outside of a compact set with $Lf \leq 0$ and $\lim_{x\uparrow\infty} f(x) = 0$.
- The process is recurrent if there exists a positive function f defined outside of a compact set with $Lf \leq 0$, $\lim_{x\uparrow\infty} f(x) = +\infty$.

Using the identity (+) with $\delta = 1$ we may reduce questions of transience and recurrence to the corresponding questions for the limiting Brownian motion process. Suppose for example that the manifold M is simply connected and complete with all sectional curvatures non-positive (defined below). A classical theorem of Cartan and Hadamard [25] asserts that M is diffeomorphic to R^d by the exponential map relative to an arbitrary reference point m. To study transience, we assume further that all sectional curvatures of M are bounded above by a constant $-k^2 < 0$. Standard comparison theorems from geometry guarantee that $\Theta'(x)/\Theta(x) \geq (d-1)k$. Let $f_o(x) = 1/r$, $r = dist(x, m)$ where m is a reference point. Then $Df_o = O(1/r^2)$, $D^2 f_o = 0(1/r^3)$, $D^3 f_o = O(1/r^4), d\Pi D^2 f_o = 2/r^3 + [\Theta'/\Theta]/r^2 \leq (1-d)(k/r^2)(1+O(1/r))$. Letting $f = f_o + Df_o + D^2 f_o$ we have from (+) that $L_1 f = \Pi D^2 f_o + D^3\ f_o \leq (1-d)(k/r^2) + O(1/r^3) < 0$ for large r while $f = O(1/r)$, $r \to \infty$. Thus we see that *the isotropic transport process on a Cartan-Hadamard manifold of negative curvature is transient.*

I.5. Curvature of Riemannian manifolds

To explore the deeper geometric structure, we introduce the sectional curvature corresponding to a two-plane $\Pi^{(2)} \subseteq M_m$. This is done by defining Brownian motion $X_t^{\Pi^{(2)}}$ on the two-dimensional surface $exp_m \Pi^{(2)} \subseteq M$. The exit time is defined by $T_\epsilon^{\Pi^{(2)}} = inf\{t \geq 0 : dist(X_t^{\Pi^{(2)}}, m) = \varepsilon\}$. When $\varepsilon \to 0$ the mean exit time $E_m(T_\epsilon^{\Pi^{(2)}}) \sim \frac{1}{2}\varepsilon^2$ and we define the *sectional curvature* by

$$K_m^{\Pi^{(2)}} = \lim_{\epsilon\downarrow 0} \frac{E_m(T_\epsilon^{\Pi^{(2)}}) - \frac{1}{2}\varepsilon^2}{\varepsilon^4}.$$

One can show that

(i) If $(M, g) = (\mathbb{R}^d, g_o)$, then $K_m^{\Pi^{(2)}} \equiv 0$.

(ii) If $(M, g) = (S^d, g_o)$, then $K_m^{\Pi^{(2)}} = const. > 0$.

(iii) If (M, g) is an imbedded surface in $\mathcal{R}^3$, then $K_m^{\Pi^{(2)}}$ is $\frac{1}{24}\times$ Gauss curvature of M at m.

Positive sectional curvature means that Brownian motion on M moves slower than its Euclidean counterpart; negative curvature means that Brownian motion on M moves faster than its Euclidean counterpart. This can be made precise in a global sense by the following "strong law of large numbers" [20]

$$(*) \qquad \lim_{t\to\infty} \frac{1}{t} dist(X_t, m) = \lim_{r\to\infty} \sqrt{-K^{\Pi}_{rw}} > 0.$$

Here M is a simply connected and complete manifold of non-positive sectional curvature. Cartan-Hadamard guarantees that M is diffeomorphic to $\mathbb{R}^d$ via the exponential mapping, so that we can use polar coordinates in the tangent space as a global coordinate system for M with $0 \leftrightarrow m \in M$. The right side of $(*)$ is the limit of the sectional curvature along the geodesic ray $t \to exp_m(tw)$, assumed to exist and be positive.

Curvature may also be used to identify Riemannian manifolds to within "local isometry". We say that (M_1, g_1) and (M_2, g_2) are *locally isometric* if each $m \in M_1$ has a neighborhood U and a map $\varphi : U \to M_2$ such that $d\varphi \circ g_1 = g_2$ on U. For example the surface of a right circular cylinder in $\mathbb{R}^3$ is locally isometric to $\mathbb{R}^2$. A fundamental theorem of Cartan, Ambrose and Hicks [25] states that if two manifolds can be mapped to one another preserving sectional curvature, then they are locally isometric. For example, if the sectional curvatures of a manifold are constant, then the manifold is locally isometric to a manifold of constant sectional curvature.

The standard simply connected manifolds of constant sectional curvature are as follows:

$$\begin{array}{lll} K = 0 & M = \mathbb{R}^d & \text{(standard Euclidean metric)} \\ K = a^2 > 0 & M = S^d(a) = \{m \in \mathbb{R}^{d+1} : |m|^2 = \frac{1}{a^2}\} & \text{(imbedded metric)} \\ K = -a^2 < 0 & M = H^d(a) = \{m \in \mathbb{R}^d : |m|^2 < \frac{4}{a^2}\} & \end{array}$$

$$\text{with the metric } (X, Y)_m = \frac{\sum_{i=1}^{d} x_i y_i}{1 - \frac{a^2 |m|^2}{4}}.$$

$H^d(a)$ is the standard hyperbolic space of constant sectional curvature, a generalization of the Poincaré plane of non-Euclidean geometry.

Closely related to the sectional curvature are the Ricci curvature and scalar curvature, obtained by "contraction". The *Ricci curvature* is a bilinear form defined on the tangent space M_m as follows: $Ric_m(0,0) = 0$; if $X \neq 0$ we set

$$Ric_m(X, X) = |X|^2 \sum_{j=2}^{d} K_m^{\Pi^{(2)}(X, e_j)}$$

where $(e_2, ..., e_d)$ is an orthonormal basis of the orthocomplement of X in M_m. (It is proved that the above definition doesn't depend on the basis chosen.) In general, we let

$$Ric_m(X, Y) = \frac{1}{4}[Ric_m(X+Y, X+Y) - Ric_m(X-Y, X-Y)].$$

In case $K^{\Pi} = k$, then $Ric_m(X,Y) = (d-1)k < X,Y >_m$, but not conversely*. If $Ric_m(X,Y) = f(m)(X,Y)_m$ for a real-valued function $f(m)$, then we say that g is an *Einstein metric*. (In this case it follows that $f(m) = const.$ if $d \geq 3$).

The *scalar curvature* is the function

$$\tau_m = \sum_{i=1}^{d} Ric_m(e_i, e_i) = \sum_{1 \leq i \neq j \leq d} K_m^{\Pi^2(e_i, e_j)}$$

where $(e_1, ..., e_d)$ is a basis of M_m and $\Pi^2(e_i, e_j)$ is the two-plane spanned by (e_i, e_j). If M has constant sectional curvature k, then $\tau_m = d(d-1)k$. The *traceless Ricci tensor* is defined by

$$Ric_m^{\#}(X,Y) = Ric_m(X,Y) - \frac{\tau_m}{d} < X,Y >_m .$$

In order to answer various inverse questions, we develop the following orthogonal decomposition of the curvature of a Riemannian manifold

$$(**) \qquad R = Weyl(R) + Ric^{\#}(R) + \tau(R).$$

Each member of this equation is a multi-linear function of four vector arguments $X, Y, Z, W \in M_m$. Proceeding from right to left, we have

$$\begin{aligned} d(d-1)\tau(R; X,Y,Z,W) &\doteq \tau_m(< X,Z >_m < Y,W >_m \\ &\quad - < X,W >_m < Y,Z >_m) \\ (d-2)Ric^{\#}(R; X,Y,Z,W) &\doteq < X,Z >_m Ric_m^{\#} < Y,W >_m \\ &\quad + < Y,W >_m Ric_m^{\#} < X,Z > \\ &\quad - < X,W >_m Ric_m^{\#} < Y,Z > \\ &\quad - < Y,Z >_m Ric_m^{\#}(X,W). \end{aligned}$$

Each of the tensors is odd under the interchange $X \leftrightarrow Y$, $Z \leftrightarrow W$, even under the interchange $(X,Y) \leftrightarrow (Z,W)$ and satisfies Jacobi's identity w.r.t. cyclic permutations of $X \to Y \to Z \to X$. The tensor $R(X,Y,Z,W)$ is uniquely defined by these symmetry properties together with the normalization that $R(e_i, e_j, e_i, e_j) = K^{\Pi^2(e_i, e_j)}$ if $i \neq j$ and $(e_1, ..., e_d)$ is an orthonormal basis of M_m. The conformal tensor $Weyl(R)$ is defined by $(**)$. This decomposition is orthogonal in the sense of the inner product $< A, B > = \sum_{i,j,k,l=1}^{d} A(e_i, e_j, e_k, e_l) B(e_i, e_j, e_k, e_l)$ in the space of 4-tensors.

The following facts are easily demonstrated:

(i) $K \equiv 0$ iff $Weyl(R) \equiv 0$, $Ric^{\#}(R) \equiv 0$, $\tau(R) \equiv 0$.

---

* The product manifold $M = S^n \times S^n$ $(n \geq 2)$ is Einstein, but doesn't have constant sectional curvature.

(ii) $K \equiv k$ iff $Weyl(R) \equiv 0$, $Ric^{\#}(R) \equiv 0$ $(d > 2)$.

(iii) g is Einstein iff $Ric^{\#}(R) \equiv 0$ $(d > 2)$.

(iv) g is conformally flat iff $Weyl(R) \equiv 0$ $(d > 3)$.

(v) If $d = 2$, then $Weyl(R) \equiv 0$, $Ric^{\#}(R) \equiv 0$.

(vi) If $d = 3$, then $Weyl(R) \equiv 0$.

The functions $|Weyl(R)|^2$, $|Ric^{\#}(R)|^2$, $\tau(R)^2$ form a basis of the quadratic curvature invariants of the curvature tensor. They enter in many asymptotic formula in Riemannian geometry. For example the volume of the ball $B_m(r)$ has the three-term asymptotic expansion of the form [18]

$$vol\ B_m(r) = \frac{(\pi r^2)^{\frac{d}{2}}}{(d/2)!}[1 - c_1 \tau_m r^2 + r^4(c_2|Weyl|^2 + c_3|Ric^{\#}|^2 + c_4\tau^2) + O(r^6)]$$

when c_1, c_2, c_3, c_4 are constants which only depend on the dimension. In the case of $M = \mathcal{R}^d$, the bracketed term is identically zero and one retrieves the classical formula for the volume of a Euclidean ball.

To develop expansions of this type, one needs to first obtain an expansion of the metric in terms of the curvature. In a neighborhood of $m \in M$, this has the form

$$g_{ij}(x) = \delta_{ij} - \frac{1}{3}\sum_{a,b=1}^{d} R_{iajb} x_a x_b + O(|x|^3)$$

where $(x_1, ..., x_d)$ is a system of Cartesian coordinates for the tangent space M_m with basis $(e_i, ..., e_d)$ and $R_{iajb} = R(e_i, e_a, e_j, e_b)$, $g_{ij}(x) = < e_i, e_j >_p$ with $p = exp_m(\sum_{i=1}^{d} x_i e_i)$.

The Laplacian of the manifold can be written in terms of the curvature by a formula of the type

$$\Delta = \Delta_{-2} + \sum_{k=0}^{\infty} \Delta_k.$$

Here $\Delta_{-2} = \sum_{i=1}^{d} \frac{\partial^2}{\partial x_i^2}$ is the Euclidean Laplacian of the tangent space M_m. The operators Δ_k are second-order differential operators with polynomical coefficients so that Δ_k increases the degree of a polynomial by k units. For example

$$\Delta_0 = \frac{1}{3}\sum_{i,a,j,b=1}^{d} R_{iajb} x_a x_b \frac{\partial^2}{\partial x_i \partial x_j} - \frac{2}{3}\sum_{i,j=1}^{d} Ric_m(e_i, e_j) x_j \frac{\partial}{\partial x_i}$$

preserves the degree of a polynomial. For the purpose of systematic application, we introduce a small parameter $\varepsilon > 0$ and a family of mappings $\Phi_\varepsilon : C^\infty(M_m) \to C^\infty(M)$ by

$$(\Phi_\varepsilon f)(exp_m x) = f(\frac{x}{\varepsilon}).$$

Then we have for each $f \in C^\infty(M_m)$, $N \in Z^+$

$$\Phi_\varepsilon^{-1} \Delta \Phi_\varepsilon f = \varepsilon^{-2} \Delta_{-2} f + \sum_{k=0}^{N} \varepsilon^k \Delta_k f + O(\varepsilon^{N+1}).$$

This will allow us to reduce problems involving Δ in a small sphere to perturbation problems in M_m, based on the Euclidean Laplacian Δ_{-2}. In the second lecture we'll discuss the exit time of Brownian motion from the ε-sphere, defined as

$$T_\varepsilon = inf\{t > 0 : dist(X_t, m) = \varepsilon\}.$$

In case $(M, g) = (\mathcal{R}^d, g_o)$, this has the following well-known properties:

(i) The P_m – law of $\frac{T_\varepsilon}{\varepsilon^2}$ doesn't depend on ε.

(ii) The P_m – law of X_{T_ε} is normalized Lebesgue measure.

(iii) X_{T_ε} and T_ε are P_m-independent.

In the case of a general Riemannian manifold, these properties hold in the limit $\varepsilon \to 0$:

(i') $\frac{T_\varepsilon}{\varepsilon^2} \overset{P_m}{\to}$ limit law $(\varepsilon \downarrow 0)$.

(ii') $X_{T_\varepsilon} \overset{P_m}{\to}$ Lebesgue measure $(\varepsilon \downarrow 0)$.

(iii') The P_m law of $(X_{T_\varepsilon}, T_\varepsilon)$ factors $(\varepsilon \downarrow 0)$.

The latter properties do not involve the gometry of (M, g). In order to bring in the geometry, we develop asymptotic expansions for the mean exit time, exit place distribution and joint laws.

Lecture II : Local stochastic differential geometry-intrinsic approach

II.1. Mean exit time of Brownian motion from a small ball

In order to develop asymptotic expansions for various quantities of interest, we recall some basic notions of stochastic analysis. Let M be a differentiable manifold of dimension d and let A be a second-order differential operator on M. A diffusion process (X_t, P_x) is said to be generated by A if $\forall f \in C^\infty(M)$, the process $M_t^f \doteq f(X_t) - \int_0^t Af(X_s)ds$ is a P_x-martingale. An immediate consequence is *Dynkin's identity*:

$$E_x f(X_T) = f(x) + E_x \int_0^T Af(X_s)ds.$$

Here T is a stopping time with $E_x(T) < \infty$. For example if A is a non-degenerate elliptic operator, we may take $T = inf\{t \geq 0 : X_t \notin U\}$ for a bounded open set U.

Dynkin's identity immediately yields the probabilistic representation of the solution of two classical boundary value problems

$$\begin{array}{lll} Af_1 = -1 \text{ in } B, & f_1 = 0 \text{ on } \partial B \Rightarrow f_1(x) = E_x(T) \\ Af_2 = 0 \text{ in } B, & f_2 = g \text{ on } \partial B \Rightarrow f_2(x) = E_x g(X_T). \end{array}$$

Here B is a bounded open set and $T = inf\{t \geq 0 : X_t \notin B\}$.

Closely related is the α-discounted version of Dynkin's identity written as

$$E_x(e^{-\alpha T} f(x_T)) = f(x) - E_x \int_0^T e^{-\alpha s}(\alpha - A)f(X_s)ds \qquad (\alpha > 0).$$

From this we conclude

$$Af_3 = \alpha f_3 \text{ in } B, \qquad f_3 = g \text{ on } \partial B \Rightarrow f_3(x) = E_x(e^{-\alpha T} g(X_T)).$$

In practice one can rarely find exact solutions of these classical problems. Nevertheless, Dynkin's identity gives the following representation of approximate solutions:

$$\begin{array}{lll} Af_\epsilon = -1 + O(\epsilon^k) \text{ in } B_\epsilon, & f_\epsilon = 0 \text{ on } \partial B_\epsilon \Rightarrow f_\epsilon(x) = E_x(T_\epsilon)(1 + O(\epsilon^k)) \\ Af_\epsilon = O(\epsilon^k) \text{ in } B_\epsilon, & f_\epsilon = g \text{ on } \partial B_\epsilon \Rightarrow f_\epsilon(x) = E_x g(X_{T_\epsilon}) + O(\epsilon^k E_x(T_\epsilon)). \end{array}$$

Here B_ϵ is a sequence of open sets with $x \in B_\epsilon$, diam $B_\epsilon \downarrow 0$.

As a first application of these methods, we let (M, g) be a d-dimensional Riemannian manifold and b a smooth vector field. The differential operator $A = \frac{1}{2}\Delta + b \cdot \nabla$ generates a diffusion process, where Δ is the Laplacian of the metric. T_ϵ is the exit time from the ball of radius ϵ:

$$T_\epsilon \doteq inf\{t > 0 : dist(X_t, m) = \epsilon\}.$$

Our first result gives the joint influence of (g, b) on the mean exit time.

Theorem 0. *For each $m \in M$ we have for $\varepsilon \downarrow 0$*

$$E_m(T_\varepsilon) = c_0\varepsilon^2 + c_1\varepsilon^4(\tau_m - 6(div\, b + |b|^2)) + 0(\varepsilon^6)$$

where c_0, c_1 are dimension constants and τ is the scalar curvature of the Riemannian metric.

This is proved by finding a function f_ε satisfying $Af_\varepsilon = -1 + O(\varepsilon^4)$ in the ball B_ε with $f_\varepsilon = 0$ on ∂B_ε. Explicitly we choose

$$f_\varepsilon(x) = \frac{\varepsilon^2 - |x|^2}{2d} + \frac{\varepsilon(\varepsilon^2 - |x|^2)}{12d^2(d+2)} \sum_{i=1}^{d} b_i x_i + \varepsilon^4 f_2(x/\varepsilon)$$

for a suitable choice of f_2. The above prescription dictates a unique choice for f_2 from which we compute $f_2(0)$ and obtain the theorem.

From this result we may determine the geometry of a surface ($d = 2$) from the mean exit time of Brownian motion.

Corollary 0. *If $(M^1, g^1), (M^2, g^2)$ are two-dimensional Riemannian manifolds such that the exit times of Brownian motion satisfy $\forall m \in M_1, m_2 \in M_2$ $E_{m_1}(T_\varepsilon^{(1)}) - E_{m_2}(T_\varepsilon^{(2)}) = O(\varepsilon^6)$ ($\varepsilon \downarrow 0$), then $\tau_{m_1}^{(1)} = \tau_{m_2}^{(2)}$, in particular the manifolds are locally isometric.*

In order to discuss inverse problems in higher dimensions one must obtain an additional term in the above asymptotic expansion. We restrict attention to standard Brownian motion ($A = \frac{1}{2}\Delta$, $b = 0$).

Theorem 1. *Let T_ε be the exit time of Brownian motion from the ball of radius ε. There exist constants $c_i = c_i(d)$, such that for each $m \in M$ we have for $\varepsilon \downarrow 0$*

$$E_m(T_\varepsilon) = c_0\varepsilon^2 + c_1\tau_m\varepsilon^4 + \varepsilon^6(c_2|Weyl|^2 + c_3|Ric^\#|^2 + c_4\tau^2 + c_5\Delta\tau)_m + O(\varepsilon^8).$$

The constants satisfy $c_i > 0$ for $i \neq 3$ and $c_3 > 0$ iff $d < 6$.

From this expansion we may retrieve the Riemannian metric from the mean exit time in low dimensions.

Corollary 1. *If (M, g) is a d-dimensional Riemannian manifold and $(\widetilde{M}, \widetilde{g})$ is a manifold of constant sectional curvature such that $\forall m \in M$, $E_m(T_\varepsilon) - \widetilde{E}_m(\widetilde{T}_\varepsilon) = O(\varepsilon^8)$, $\varepsilon \downarrow 0$, then (M, g) is locally isometric to $(\widetilde{M}, \widetilde{g})$ provided that $d = dim\, M < 6$.*

In dimension 6 we have the following example.

Example. Let $M = S^3(1) \times H^3(1)$ with the product metric, then $E_m(T_\varepsilon) = c_0\varepsilon^2 + O(\varepsilon^{10})$, $\varepsilon \downarrow 0$. In other words M has the same expansion as for the case of Euclidean $\mathbb{R}^6$.

To prove theorem 1 we construct a trial solution in the form $f_\varepsilon = \Phi_\varepsilon(\varepsilon^2 F_0 + \varepsilon^4 F_2 + \varepsilon^5 F_3 + \varepsilon^6 F_4)$ where F_0, F_2, F_3, F_4 are obtained by solving the following boundary-value problems

$$\begin{array}{ll} \Delta_{-2} F_0 + 1 = 0 \text{ in } \bar{B}_1, & F_0 = 0 \text{ on } \partial\bar{B}_1 \\ \Delta_{-2} F_2 + \Delta_0 F_0 = 0 \text{ in } \bar{B}_1, & F_2 = 0 \text{ on } \partial\bar{B}_1 \\ \Delta_{-2} F_3 + \Delta_1 F_0 = 0 \text{ in } \bar{B}_1, & F_3 = 0 \text{ on } \partial\bar{B}_1 \\ \Delta_{-2} F_4 + \Delta_0 F_2 + \Delta_2 F_0 = 0 \text{ in } \bar{B}_1, & F_4 = 0 \text{ on } \partial\bar{B}_1 . \end{array}$$

It is immediate that $\Delta f_\varepsilon = -1 + O(\varepsilon^8)$, $\varepsilon \downarrow 0$. It remains to find $F_2(0)$, $F_3(0)$, $F_4(0)$. Clearly $F_0(x) = \frac{1-|x|^2}{2d}$, thus $F_0(0) = \frac{1}{2d}$. To determine the higher terms, we appeal to the Pizetti formula: $\Delta_{-2} F + g = 0$ in B_1, $F = 0$ on $\partial B_1 \Rightarrow F(0) = -\sum_{j\geq 1} \frac{1}{2^j j! d(d+2)\ldots(d+2j-2)} \Delta_{-2}^{j-1} g$. For the choice $g = \Delta_0 f_0$ we have $\Delta_{-2}\Delta_0 f_0 = \Delta_0 \Delta_{-2} f_0 - \frac{2}{3}(Ric)_{jk} \frac{\partial^2 f_0}{\partial x_j \partial x_k} = -\frac{2}{3d}(Ric)_{jk}\delta_{jk} = -\frac{2\tau_m}{3d}$, leading to $F_2(0) = \frac{\tau_m}{12d^2(d+2)}$. To compute $F_4(0)$, we must first obtain F_2 in the form $F_2(x) = e_1(1-|x|^2) + e_2 Ric(x,x)(1-|x|^2)$ for suitable constants e_1, e_2. One then computes $\Delta_0 F_2 + \Delta_2 F_0$ and applies the Pizetti formula to compute $F_4(0) = c_2|Weyl|^2 + c_3|Ric^{\#}|^2 + c_4\tau^2 + c_5\Delta\tau$. The details are carried out in [6].

II.2. Exit distribution of Brownian motion from a small ball

In order to study the exit place X_{T_ε}, we map the problem to the unit sphere by the following transformation. Let $f \in C^\infty(S^{d-1})$, a smooth function on the unit sphere. We set

$$\begin{aligned} H_\varepsilon f(p) &= E_p\{(\Phi_\varepsilon f)(X_{T_\varepsilon})\} \\ &= \int_{S^{d-1}} f(\theta) h_\varepsilon(p, d\theta) \end{aligned}$$

$h_\varepsilon(p,\cdot)$ is the *harmonic measure*, giving the probability distribution of the exit place X_{T_ε}. It is immediate that $u \doteq H_\varepsilon f$ is the solution of $\Delta u = 0$, in B_ε with $u = \Phi_\varepsilon f$ on ∂B_ε. We find an approximate solution to within $O(\varepsilon^4)$ and compute $u(p)$ to obtain the following result.

Theorem. [11] $H_\varepsilon f(m) = \int_{S^{d-1}} f(\theta)\,[1 - \frac{\varepsilon^2}{12} Ric^{\#}_m(\theta,\theta)]\,\omega_m(d\theta) + O(\varepsilon^3)$, $\quad \varepsilon \downarrow 0$.

From this we immediately obtain the following corollary.

Corollary. *If* $\forall m \in M$, $H_\varepsilon f(m) = \int_{S^{d-1}} f(\theta)\omega_m(d\theta) + O(\varepsilon^3)$ *then* (M,g) *is an Einstein manifold.*

We indicate the computation of the quadratic term. We look for a trial solution in the form $F = \Phi_\varepsilon(f_0 + \varepsilon^2 f_2)$ where

$$\begin{array}{ll} \Delta_{-2} f_0 = 0 \text{ in } \bar{B}_1, & f_0 = f \text{ on } \partial\bar{B}_1 \\ \Delta_{-2} f_2 + \Delta_0 f_0 = 0 \text{ in } \bar{B}_1, & f_2 = 0 \text{ on } \partial\bar{B}_1 \\ \Delta_{-2} f_3 + \Delta_1 f_0 = 0 \text{ in } \bar{B}_1, & f_3 = 0 \text{ on } \partial\bar{B}_1 . \end{array}$$

It follows that $\Delta F = O(\varepsilon), E_m(\Phi_\varepsilon f) = E_m F(x_T) = F(m) + O(\varepsilon^3)$. It remains to compute $F(m) = f_0(0) + \varepsilon^2 f_2(0)$; from classical potential theory we have $f_0 = \int_{S^{d-1}} \frac{1-|x|^2}{|x-z|^n} f(z)\omega(dz)$. Thus $f_0(0) = \int_{S^{d-1}} f(z)\omega(dz)$. To compute $f_2(0)$ we note that the classical Pizetti formula [11] gives the expansion $f_2(0) = -\sum_{j\geq 1} \frac{1}{2^j j!(d)(d+2)\ldots(d+2j-2)} \Delta_{-2}^{j-1} \Delta_0 f_0(0)$. It remains to compute

$$\Delta_0 f_0(0) = 0.$$

$$\Delta_{-2}\Delta_0 f_0 = \Delta_0\Delta_{-2} f_0 + \frac{2}{3}(Ric)_{jk} \frac{\partial^2 f_0}{\partial x_j \partial x_k} = -\frac{2}{3}(Ric)_{jk} \frac{\partial^2 f}{\partial x_j \partial x_k}.$$

$$\Delta_{-2}^2(\Delta_0 f_0) = \Delta_0(\Delta_{-2}^2 f_0) - \frac{4}{3}(Ric)_{jk} \frac{\partial^n}{\partial x_j \partial x_n}(\Delta_{-2} f) = 0.$$

The higher powers are zero. Finally we have from the Poisson formula

$$\frac{\partial^2 f_0}{\partial x_j \partial x_k}(0) = \int_{S^{d-1}} [d(d+2) z_j z_k - (d+2)\delta_{jk}] f(z)\omega(dz).$$

$$\frac{2}{3}(Ric)_{jk} \frac{\partial^2 f}{\partial x_j \partial x_k}(0) = \frac{2}{3} d(d+2) \int_{S^{d-1}} Ric^{\#}(z,z) f(z)\omega(dz).$$

$$f_2(0) = -\frac{1}{12}\int_{S^{d-1}} Ric^{\#}(z,z) f(z)\omega(dz).$$

This is the required formula.

The Theorem may be used to give meaning to the motion of "most probable direction" of Brownian motion leaving a point. Suppose that the Ricci tensor has a unique eigenvector ν_0 with smallest eigenvalue : $Ric_m^{\#}(\nu_0, \nu_0) < Ric_m^{\#}(\nu, \nu)\ \forall \nu \in S_m$. Then $\exists \delta > 0$ such that $P_m(|\Phi_\varepsilon^{-1}(X_{T_\varepsilon}) - \nu_0| < \delta) > P_m(|\Phi_\varepsilon^{-1}(X_{T_\varepsilon}) - \nu| < \delta)$ for sufficiently small $\varepsilon > 0$.

II.3. Independence of exit time and exit place

To study the joint law of $(T_\varepsilon, X_{T_\varepsilon})$ we introduce the Laplace transform

$$u(p) = E_p(e^{-\alpha(\frac{T_\varepsilon}{\varepsilon^2})}(\Phi_\varepsilon f)(X_{T_\varepsilon}))$$

which is the classical solution of $(\frac{\alpha}{\varepsilon^2} - \Delta)u = 0$ in B_ε with $u = \Phi_\varepsilon f$ on ∂B_ε. An approximate solution is written in the form $u = \Phi_\varepsilon(u_0 + \varepsilon^2 u_1 + \varepsilon^3 u_2)$. We have the following result.

Theorem.

$$u_\varepsilon(m) = \int_{S^{d-1}} [c_0(\alpha) + \varepsilon^2(c_1(\alpha) Ric_m^{\#}(\theta) + c_2(\alpha)\tau_m) + \varepsilon^3(c_3(\alpha)\partial_\theta Ric_m^{\#}(\theta,\theta) + c_4(\alpha)\partial_\theta \tau_m)] f(\theta)\omega_m(d\theta) + O(\varepsilon^4)$$

when $c_0(\alpha), \ldots, c_4(\alpha)$ are constants which depend on the dimension.

We can use this to compute the difference

$$E_m(e^{-\alpha(\frac{T_\varepsilon}{\varepsilon^2})}(\Phi_\varepsilon f)(X_{T_\varepsilon})) - E_m(e^{-\alpha(\frac{T_\varepsilon}{\varepsilon^2})})E_m\{(\Phi_\varepsilon f)(X_{T_\varepsilon})\}$$
$$= \varepsilon^3 c_4(\alpha) \int_{S^{d-1}} \partial_\theta \tau_m f(\theta)\omega_m(d\theta) + O(\varepsilon^4)$$

and obtain the following corollary.

Corollary. *Suppose that $\forall m \in M$, the random variables $T_\varepsilon, X_{T_\varepsilon}$ are independent. Then the scalar curvature is constant.*

In order to obtain a stronger conclusion, H. R. Hughes [13] added the hypothesis of nonzero scalar curvature, to obtain the following rsult.

Proposition. *Suppose that every $\varepsilon > 0$, $m \in M$ the random variables T_ε and X_{T_ε} are independent and that $\tau_m \neq 0 \; \forall \; m \in M$. Then $Ric^\#_m \equiv 0$, i.e. g is an Einstein metric.*

H. R. Hugues also proved that a condition of "super-Einstein" holds (see [13], p.27 ff).

II.4. Large deviations and the first Dirichlet eigenvalue

The Laplacian of the manifold is a formally self-adjoint operator for which we may consider the eigenvalue problem $\Delta u + \lambda u = 0$ in B_ε with $u = 0$ on ∂B_ε. Classical estimates guarantees a discrete spectrum of eigenvalues $0 < \lambda_1 < \lambda_2 \leq \ldots$ with $\lim_{n\to\infty} \lambda_n = +\infty$, and corresponding eigenfunctions φ_n. The heat equation $\frac{\partial u}{\partial t} = \Delta u$ with zero boundary conditions may be solved by an eigenfunction series $u = \sum_{n\geq 1} < f, \phi_n > \phi_n e^{-\lambda_n t}$ when $f(x) = u(0^+, x)$. For the case $f = 1$ we obtain from Dynkin's formula $u = P_m(T_\varepsilon > t) = const \; e^{-\lambda_1(m,\varepsilon)t} + O(e^{-t\lambda_2(m,\varepsilon)})$, $\varepsilon \downarrow 0$. Therefore the principal eigenvalue $\lambda_1(m, \varepsilon)$ characterizes the large deviation asymptotics of the principal eigenvalue according to

$$P_m(T_\varepsilon > t) \cong e^{-t\lambda_1(\varepsilon,m)} \qquad (\varepsilon \downarrow 0).$$

In order to obtain an asymptotic expansion for $\lambda_1(\varepsilon, m)$ we may employ the general method used to analyse the mean exit time, to obtain the following result.

Theorem. *When $\varepsilon \downarrow 0$ we have the expansion*

$$\lambda_1(\varepsilon, m) = \frac{c_0}{\varepsilon^2} + c_1\tau_m + c_2[|R|^2 - |Ric|^2 + 6\Delta\tau]_m + O(\varepsilon^4)$$

where c_0, c_1, c_2 depend only on the dimension and satisfy $c_0 > 0$, $c_1 < 0$, $c_2 < 0$.

From this we have the following corollaries.

Corollary 1. *Let $d = \dim M < 6$. If for every $m \in M$ we have $\lambda_1(\varepsilon, m) = c_0/\varepsilon^2 + O(\varepsilon^4)$, then (M, g) is locally isometric to R^d with the Euclidean metric.*

Corollary 2. *If for every $m \in M$ we have $\lambda_1(\varepsilon, m) = c_0/\varepsilon^2 + O(\varepsilon^4)$ and the Ricci tensor is semidefinite* (esp. if g is an Einstein metric) *then (M, g) is locally isometric to R^d with the Euclidean metric.*

The proof is effected by means of an approximate solution to the eigenvalue problem $\Delta f_{(\epsilon)} + \lambda_{(\epsilon)} f_{(\epsilon)} = 0$ where $\sum_{j=0}^{k+2} (\Delta_{k-j} + \lambda_{k-j}) f_j = 0$ for $k = -2, 1, 0, 1, 2$ and $f_j = 0$ on ∂B_ϵ with $\|f_0\| = 1$ and $(f_m, f_0) = 0$ for $k > 0$ where the inner product is computed in $L^2(B_1)$ with respect to Lebesgue measure. Explicit calculation, carried out in [15] give the non-zero values of $\lambda_{-2}, \lambda_0, \lambda_2$ leading to the result. We can paraphrase the corollaries as giving conditions under which we can retrieve the structure of the Riemannian metric from the large derivation asymptotics of the first exit time from small balls. Further results are given in [15].

II.5. Stochastic parallel transport of conditional Brownian motion

By generalizing the construction of the isotropic transport process, we can construct a process of "stochastic parallel transport" on the tangent bundle of any Riemannian manifold. This coincides with the construction of Itô and Malliavin as described by Ikeda and Watanabe [1].

Let the tangent bundle be written as

$$TM = \{(m, \xi) : m \in M, \xi \in M_m\}.$$

We will construct a canonical Markov process (isotropic transport process) on the bundle of 2-frames

$$T^{(2)}(M) = \{(m, \xi, \eta) : m \in M,\ \xi \in M_m,\ \eta \in M_m\}.$$

To do this, let $\tilde{D}$ be the *canonical horizontal vector field* on $T^{(2)}(M)$ defined by

$$\tilde{D} f(m, \xi, \eta) = \frac{d}{dt} f(\gamma(t), \xi(t), \gamma'(t))|_{t=0}$$

where $\gamma(t)$ is the geodesic with $\gamma(0) = m, \gamma'(0) = \eta \in M_m$ and $\xi(t)$ is the parallel transport of the vector $\xi = \xi(0)$ along the geodesic $\{\gamma(t), t \geq 0\}$. In local coordinates, we may write

$$\tilde{D} f(m, \xi, \eta) = \sum_{i=1}^{d} \xi_i \frac{\partial f}{\partial x_i} - \sum_{i,j,k=1}^{d} (\Gamma_{ij}^k(m) \eta_i \eta_j \frac{\partial f}{\partial \eta_k} - \Gamma_{i,j}^k(m) \xi_i \eta_j \frac{\partial f}{\partial \xi_k}).$$

A *canonical projection operator* Π is defined exactly as in I.1 by

$$\Pi f(m,\xi) = \int_{S^{d-1}} f(m,\xi,\eta)\omega_m(d\eta)$$

when the integration is with respect to the uniform distribution on the unit sphere S^{d-1}. These operators satisfy the properties that $\Pi^2 = \Pi, \Pi 1 = 1$ and $\Pi\tilde{D}\Pi = 0$. If $\delta > 0$ is a small parameter, we may generate an isotropic transport process by means of the operator

$$L_\delta = \frac{\tilde{D}}{\delta} + \frac{\Pi - I}{\delta^2}.$$

This is a Markov process $(X_\delta(t), \xi_\delta(t), \eta_\delta(t))$ on $T^{(2)}(M)$ which moves the vector $\xi(0)$ along piecewise geodesic paths for exponential times, choosing independent directions according to the uniform distribution on the tangent sphere. As before, we may verify the identity $L_\delta(f + \delta\tilde{D}f + \delta^2\tilde{D}f) = \Pi\tilde{D}^2 f + \delta D^3 f$ and use standard methods of weak convergence to show that the sequence of processes on $T^{(2)}M$ converges weakly to a diffusion process generated by the *horizontal Laplacian*

$$L = \Pi\tilde{D}^2.$$

This is a degenerate diffusion process on TM whose projection onto the base manifold M is precisely the Brownian motion discussed in I.3. Following previous authors, we use the term *stochastic parallel transport.*

Having constructed the stochastic parallel transport process, it is natural to develop the methods of II.1 - II.3 to solve problems on TM. For this purpose we need to extend the homogeneous decomposition of the Laplacian, in the form

$$\Phi_\epsilon^{-1} L \Phi_\epsilon \cong d^{-1}[\varepsilon^{-2}\Delta_{-2} + \sum_0^\infty \varepsilon^h L_k].$$

The first two new operators are given by

$$L_0 = \Delta_0 - R_{pajh} x^a \xi_p \frac{\partial^2}{\partial x_j \partial \xi_h},$$

$$L_1 = \Delta_1 - \frac{2}{3}\partial_a R_{pbjh} x^a x^b \xi^p \frac{\partial^2}{\partial x_j \partial \xi_h} - \frac{1}{3}\partial_k R_{pabi} x^a \xi^p \frac{\partial}{\partial \xi_i},$$

where we sum over repeated indices. We can apply this to solve a boundary-value problem of the form $Lu = 0$ in B_ϵ with $u = \Phi_\epsilon(f)$ prescribed on ∂B_ϵ. This is written in the form $u = \Phi_\epsilon(W_0 + \varepsilon^2 W_2 + \varepsilon^3 W_3)$ where

$$\begin{aligned}
\Delta_{-2}W_0 &= 0 \text{ in } \bar{B}_1, & W_0 &= f \text{ on } \partial\bar{B}_1\\
\Delta_{-2}W_2 + L_0 W_0 &= 0 \text{ in } \bar{B}_1, & W_2 &= 0 \text{ on } \partial\bar{B}_1\\
\Delta_{-2}W_3 + L_1 W_0 &= 0 \text{ in } \bar{B}_1, & W_3 &= 0 \text{ on } \partial\bar{B}_1.
\end{aligned}$$

In particular, the choice $f(x,\xi) = \xi^0 g(x)$ leads to the asymptotic result [17]

$$E_{m,\xi}\{\xi_T^i \Phi_\epsilon(g(x_T))\} = \xi^i E\{\Phi_\epsilon(g(x_T))\} + \varepsilon^3 \int_{S^{d-1}} A(z,\xi) g(z) + O(\varepsilon^4)$$

where

$$A_i(z,\xi) = \frac{1}{12d(d+2)} \xi^p z^j (\partial_i R_{pj} - \partial_p R_{ij}).$$

Applying this to a function g which is concentrated near the point $z \in S_m$ and re-normalizing, we obtain the following result on "conditional stochastic parallel transport" [17].

Theorem. *For each $z \in S_m$, $\xi \in S_m$ we have for $\varepsilon \downarrow 0$*

$$E_{m,\xi}\{\xi_T^i | exp_m^{-1}(X_T) = \varepsilon z\} = \xi^i - \varepsilon^3 A_i(z,\xi) + O(\varepsilon^4).$$

Corollary. *If for $\forall m \in M$, $\varepsilon > 0$, $z \in M_m$ we have $E_{m,z}(\xi_T | exp_m^{-1}(X_T) = \varepsilon z) = \xi$, then M has constant scalar curvature.*

Lecture III : Local stochastic differential geometry-extrinsic approach

III.1. Mean exit time of Brownian motion from a tube

In the first two lectures we developed local properties of Brownian motion on an abstractly presented Riemannian manifold. In the following sections we consider inverse problems in the context of *extrinsic geometry*, where we are concerned with the imbedding of a manifold as a submanifold of a given higher-dimensional manifold.

A classical topic in this area is the Weyl tube formula and its generalizations [27]. These formulas give explicit expression to the fact that the volume of a tubular neighborhood of a submanifold depends only on the intrinsic geometry of the submanifold and the width of the neighborhood - and on no other details of the imbedding

To formulate a stochastic counterpart of the volume, we consider the Brownian motion (X_t, P_x) of a Riemannian manifold with $X_0 = m \in V$, a compact submanifold of dimension $q < d = dim\ M$. The exit time from the tubular neighborhood is defined by

$$T_\varepsilon = inf\{t > 0 : dist(X_t, V) = \varepsilon\}.$$

Here we take Brownian motion generated by Δ. The mean exit time $E_m(T_\varepsilon)$ may be integrated over the submanifold V to obtain a stochastic counterpart of the volume of the tubular neighborhood. In order to obtain an asymptotic expansion we recall Dynkin's formula $E_p(f(X_T)) - f(x) = E_p(\int_0^T \Delta f(X_s)ds)$ where $p \in M$, Δ is the Laplacian of (M, g) and T is any stopping time with $E_p(T) < \infty$. In particular the mean exit time is the solution u of $\Delta u = -1$ in the tubular neighborhood $V^{(\varepsilon)} = \{x \in M : dist(x, V) < \varepsilon\}$ with $f = 0$ on $\partial V^{(\varepsilon)}$. To obtain a concrete result we specialize further to the case of a hypersurface : $q = d - 1$. In this case we can take a system of Fermi coordinates (t, p) and decompose the Laplacian in the form [8]

$$\Delta = \frac{\partial^2}{\partial t^2} + \frac{\Theta_t}{\Theta}\frac{\partial}{\partial t} + \Delta_V$$

when $\Theta = \Theta(t, p)$ is the volume form defined by the exponential mapping and Δ_V is an operator in $V^{(\varepsilon)}$ which reduces to the natural Laplacian of V when $\varepsilon = 0$. Using the known power series expansion of $\Theta(t, p) \cong \sum_0^\infty c_n(p)t^n$, we obtain the following result [8].

Theorem. *For any hypersurface $V \subseteq M$ we have for $\varepsilon \downarrow 0$, $m \in V$*

$$E_m(T_\varepsilon) = \frac{1}{2}\varepsilon^2 + \frac{\varepsilon^4}{24}\{2\sum_1^{d-1} k_i^2 - (\sum_1^{d-1} k_i)^2 + 2\rho_N\} + O(\varepsilon^6)$$

where $(k_1, ..., k_{d-1})$ are the principal curvatures of the imbedding and $\rho_N = Ric(N, N)$ is the Ricci curvature of M in the normal direction.

From this we may deduce the stochastic counterpart of the Weyl tube formula by integrating over the compact hypersurface. We illustrate the result for surfaces in $\mathcal{R}^3$.

Corollary. *If $V \subseteq \mathcal{R}^3$ is a compact surface without boundary then $E_m(T_\epsilon) = \frac{1}{2}\epsilon^2 + \frac{\epsilon^4}{6}(H^2 - K)_m + O(\epsilon^6)$ where $H = \frac{1}{2}(k_1 + k_2)$ is the mean curvature of the imbedding and $K = k_1 k_2$ is the Gauss curvature of V.*

A fundamental fact of classical differential geometry asserts that $K = k_1 k_2$ is intrinsic to V, whereas $H = \frac{1}{2}(k_1 + k_2)$ is not intrinsic. Both terms persist when we integrate over V and thus we see that, unlike the volume, the *integrated mean exit time depends on the imbedding.*

We may also use the theorem to study inverse questions, to characterize the imbedding from the integrated mean exit time. We have the following result in $q = 2$, $d = 3$.

Corollary. *If $V \subseteq \mathcal{R}^3$ is a compact surface without boundary such that*

$$\int_V E_m(T_\epsilon) d\, vol = \frac{1}{2}\epsilon^2 vol(V) + O(\epsilon^6)$$

then V is a round sphere.

Indeed the hypothesis implies, in view of the theorem, that $0 = H^2 - K = \frac{1}{4}(k_1 - k_2)^2$ and hence $k_1 \equiv k_2$, i.e. every point is umbilic. A standard result from geometry [26] guarantees that V is a sphere. However we annot characterize the radius of the sphere from the mean exit time, since an exact calculation shown that if $V = S^2(R)$ is any sphere in $\mathcal{R}^3$, then $E_m(T_\epsilon) = \frac{1}{2}\epsilon^2$ for any $\epsilon > 0$, $m \in V$. One may ask if additional information about the distribution can be used to detect the radius. An explicit calculation [8] reveals that the probability law of T_ϵ is also independent of R; in fact it is exactly the same as the exit time distribution of one dimensional Brownian motion from the interval $(R-\epsilon, R+\epsilon)$ starting at the center.

III.2. First Dirichlet eigenvalue of a tubular neighborhood

Closely related to the mean exit time is the principal Dirichlet eigenvalue of the tubular neighborhood of width 2ϵ. We have the following result [16].

Theorem. *Let $V \subseteq \mathcal{R}^d$ be a closed imbedded hypersurface. The principal eigenvalue of the ϵ-tubular neighborhood satisfies*

$$\lambda(\epsilon, V) = \frac{\pi^2}{4\epsilon^2} + (4|V|)^{-1} \int_V \{(\sum k_i)^2 - 2\sum k_i^2\} d\, vol + O(\epsilon)$$

where $|V|$ = volume of V and k_i is the $\mathrm{i^{th}}$ principal curvature of the imbedding.

We note that the quadratic curvature invariant which appears here is exactly the same as found in III.1 for the mean exit time. Therefore we easily obtain a characterization of imbedded spheres, as follows.

Corollary. *Let V be a closed imbedded surface in $\mathcal{R}^3$. If the principal eigenvalue of the tubular neighborhood satisfies $\lambda(\varepsilon, V) = \pi^2/4\varepsilon^2 + O(\varepsilon), \varepsilon \downarrow 0$, then V is a round sphere.*

III.3. Formulation of extrinsic Brownian motion

In the first two lectures we studied Brownian motion on an abstract Riemannian manifold (M, g) without any explicit imbeddings. In III.1 - 2, we investigated some questions with respect to imbedded submanifolds. In the final three sections we explicitly address the question of Brownian motion and related eigenvalue problems on manifolds M which are given as submanifolds of $\mathcal{R}^N$. In this case we have, in addition to the Riemannian structure of M, the structure of the imbedding. In particular we may consider "extrinsic balls" with respect to the ambient metric and various invariants of the second fundamental form of the imbedding.

Brownian motion on imbedded hypersurfaces was explicitly treated by Lewis and Van den Berg [4] which we now review. On an open set $W \subseteq \mathcal{R}^N$ we suppose given a smooth function $\Psi : W \to \mathcal{R}$ with $\nabla\Psi(x) \neq 0$ for $x \in W$. The hypersurface is represented by

$$M = \{x \in \mathcal{R}^N : \Psi(x) = 0\}.$$

Let $n(x)$ be a unit normal vector field on M, assumed orientable. The projection operator $P(x)$ maps $\mathcal{R}^N \ni v \to P(x)v = v - \langle v, n \rangle n$. If $f \in C^\infty(W; \mathcal{R})$ is a smooth function, the covariant derivative is

$$\nabla f \doteq P(x)(Df).$$

This vector is tangent to M and only depends on $f|_M$. The Laplacian is written as

$$\Delta f \doteq trace\{P(x)DP(x)Df\}.$$

This may be rendered more explicit by writing

$$P(x)Df = Df - n \langle Df, n \rangle$$
$$DP(x)Df = D^2 f - Dn \langle Df, n \rangle - nD \langle Df, n \rangle .$$

But $Pn = 0$. So we obtain

$$P(x)D(P(x)Df) = P(x)D^2 f - P(Dn) \langle Df, n \rangle .$$

Recalling the coordinate form $P_{ij}(x) = \delta_{ij} - n_i(x)n_j(x)$, we have

$$trace(P(x)D^2 f) = \sum_{i,j=1}^{N} (\delta_{ij} - n_i(x)n_j(x)) \frac{\partial^2 f}{\partial x_i \partial x_j}$$

$$trace(P(x)Dn) = \sum_{i,j=1}^{N} (\delta_{ij} - n_i(x)n_j(x)) \frac{\partial n_i}{\partial x_j} = div\, n$$

with the result

$$\Delta f = \sum_{i,j=1}^{N} (\delta_{ij} - n_i(x)n_j(x)) \frac{\partial^2 f}{\partial x_i \partial x_j} - (div\ n)(x) \sum_{i=1}^{N} n_i(x) \frac{\partial f}{\partial x_i}.$$

It may be shown that this agrees with the intrinsic Laplacian introduced in I.3. To obtain the Brownian motion of the imbedded hypersurface $V \subseteq \mathcal{R}^N$ we write the following system of stochastic differential equations

$$\begin{aligned} dX_t^i &= \sum_{j=1}^{N} [\delta_{ij} - n_i(X_t)n_j(X_t)] dB_t^j - \frac{1}{2} div\ n(X_t) n_i(X_t) dt \\ &= \sum_{j=1}^{N} [\delta_{ij} - n_i(X_t)n_j(X_t)] \circ dB_t^i. \end{aligned}$$

The first is a stochastic differential equation of Itô-type. The second is a stochastic differential equation of Stratonovich type, which may be easily seen to be equivalent to the Itô equation. Furthermore it may be shown using the chain rule (= Itô formula) that $d\Psi(X_t) \equiv 0$, i.e. the solution remains on the manifold V for all time provided that $X_0 \in V$.

III.4. Mean exit time of extrinsic Brownian motion

These consideration may be used to study the *extrinsic mean exit time*, defined by

$$\tilde{T}_\epsilon = inf\{t > 0 : |X_t - m| = \epsilon\}.$$

Clearly $\tilde{T}_\epsilon \geq T_\epsilon$, the intrinsic exit time introduced in II.1. If V is a hyperplane, the two coincide. If V is a minimal imbedding ($div\ n \equiv 0$) then the mean exit time $E_m(\tilde{T}_\epsilon)$ is the same as for a hyperplane : $E_m(\tilde{T}_\epsilon) \equiv \frac{1}{2}\epsilon^2$, as one can see by explicitly solving the equation $\Delta f = -1$. In order to study inverse questions, we develop the following asymptotic expansion [9].

Theorem. *For any hypersurface $V \subseteq \mathcal{R}^N$ we have for $\epsilon \downarrow 0$*

$$E_m(\tilde{T}_\epsilon) = \frac{\epsilon^2}{2(N-1)} + \frac{\epsilon^4}{8(N+1)} H^2 + O(\epsilon^5)$$

where $H = div\ n\ /(N-1)$ is the mean curvature.

Corollary. *If for all $m \in V$ we have $E_m(\tilde{T}_\epsilon) = \frac{\epsilon^2}{2(N-1)} + O(\epsilon^5)$, then V is a minimally imbedded hypersurface.*

The proof of the theorem relies on the construction of the approximate solution $f(x) = (\varepsilon^2 - |x|^2)(a_0 + \sum a_i x_i + \sum a_{ij} x_i x_j)$ which solves $\Delta f = -1 + O(\varepsilon^3)$. The constants a_0, a_i, a_{ij} are determined up to proportionality by the equation. The exact constants can be determined by computing $E_m(T_\varepsilon)$ for a known example, such as $V = S^{N-1}(R) \subseteq R^N$, leading to the stated values. We may characterize the imbedded hyperplane by the equality of the extrinsic and intrinsic mean exit times.

Proposition. *Suppose that for all $m \in V, E_m(T_\varepsilon) = E_m(\tilde{T}_\varepsilon)$. Then V is an imbedded hyperplane.*

This is proved by the maximum principle for elliptic second-order equations [9] without reference to the asymptotic expansion.

III.5. Volume of extrinsic balls

We close this series of lectures by application of the above ideas to a non-stochastic problem: the volume of a small extrinsic ball of an imbedded submanifold $V \subseteq \mathcal{R}^N$. The volume of intrinsic balls (defined by the geodesic distance) have been thoroughly investigated in the monumental article of Gray and Van Hecke [18], who obtained four non-zero terms in the asymptotic expansion.

To formulate our result, let V^p be a p-dimensional submanifold of $\mathcal{R}^N$ and consider the ambient ball

$$\tilde{B}_\varepsilon(x_0) = \{y \in \mathcal{R}^N : |y - x_0| < \varepsilon\} \cap V^p.$$

The asymptotics of the volume of $\tilde{B}_\varepsilon(x_0)$ are governed by the extrinsic curvatures of V^p as derived from the second fundamental form $B_{x_0}(v, w) = (\Delta_V W)^\perp$ where $v, w \in V_{x_0}$ and V, W are smooth extensions of v, w; ∇ is the standard connexion of $\mathcal{R}^N$ and $\perp$ denotes the normal component. The mean curvature normal is the vector $H_{x_0} \doteq trace\, B_{x_0} = \sum_{i=1}^{p} B_{x_0}(e_i, e_i)$ where (e_i) is an orthonormal frame for V_{x_0}. The principal curvatures (k_i) are the eigenvalues of $B(e_i, e_j)$ and $\|B\|^2 = \sum_i k_i^2, \|H\| \doteq \sum_i k_i$ [26]. We have the result [19], where σ_{p-1} is the volume of the Euclidean unit ball.

Theorem. *The volume of the extrinsic neighborhood $\tilde{B}_\varepsilon(x_0)$ has the asymptotic expression when $\varepsilon \downarrow 0$*

$$vol(\tilde{B}_\varepsilon(x_0)) = \sigma_{p-1}\varepsilon^p/p + \sigma_{p-1}/8p(p+2)[2\|B\|^2 - \|H\|^2]\varepsilon^{p+2} + O(\varepsilon^{p+3}).$$

Corollary. *If V is compact surface in $\mathcal{R}^3$ such that $vol(\tilde{B}_\varepsilon(x_0)) = \pi\varepsilon^2 + O(\varepsilon^5)$, then V is a round sphere.*

Further converse theorems are obtained in [19].

Added-in remark: On the basis of some recent results of H. R. Huges (obtained after the conference), we now know that the term $O(\epsilon^{10})$ in the example following Corollary 1 of II.1 is identically zero. This answers negatively the question "Can you feel the shape of a manifold with Brownian motion?" in the case of dimension 6.

References

Definition of Brownian motion on manifolds

[1] N. Ikeda & S. Watanabe, *Stochastic differential equations and diffusion processes*, 2nd ed., North Holland, 1989.

[2] M. Pinsky, Isotropic transport process on a Riemannian manifold, *Trans. Amer. Math. Soc.*, **218** (1976), 353-360.

[3] M. Pinsky, Homogenization in stochastic differential geometry, *Publ. Res. Inst. Math. Sci., Kyoto University*, **17** (1981), 235-244.

[4] J. Lewis & N. Van den Berg, Brownian motion on a hypersurface in R^d, *Bull. London Math. Soc.*, **17** (1985), 144-150.

Exit time of Brownian motion

[5] A. Debiard, B. Gaveau & E. Mazet, Theoremes de comparison en geometrie Riemannienne, *Publ. Res. Inst. Math. Sci., Kyoto University*, **12** (1976), 391-425.

[6] A. Gray & M. Pinsky, Mean exit time from a small geodesic ball in a Riemannian manifold, *Bull. Sci. Math.*, **107** (1983), 345-370.

[7] M. Pinsky, Brownian motion in a small geodesic ball, in *"Asterisque"* Vol **132**, Actes du Colloque Laurent Schwartz (1985), 89-101.

[8] A. Gray, L. Karp & M. Pinsky, Mean exit time from a tube in a Riemannian manifold, *Probability and Harmonic Analysis*, Marcel Dekker (1986), 113-138.

[9] L. Karp & M. Pinsky, Mean exit time from an extrinsic ball, Pitman Lecture Notes, Vol **150** *"From local times to global geometry, control and physics"*, ed. K. D. Elworthy, (1986), 214-221.

Exit distribution of Brownian motion

[10] M. Pinsky, On non-euclidean harmonic measure, *Ann. Inst. H. Poincaré*, **21** (1985), 39-46.

[11] Ming Liao, Hitting distributions of small geodesic spheres, *Ann. Probab.*, **16** (1988), 1039-1050.

[12] M. Kozaki & Y. Ogura, On the independence of exit time and exit position from small geodesic balls in Riemannian manifolds, *Math. Z.*, **197** (1988), 561-582.

[13] H. R. Hughes, Hitting time and place to small geodesic spheres on Riemannian mainfolds, Ph.D. dissertation, Northwestern University, August 1988.

First eigenvalue of a small ball or tube

[14] A. Levy-Bruhl, Invariants infinitesimaux,*C. R. Acad. Sci. Paris*, Ser A-B **279** (1974), 197-200.

[15] L. Karp & M. Pinsky, The first eigenvalue of a small geodesic ball in a Riemannian manifold, *Bull. Sci. Math.*, **111** (1987), 229-239.

[16] L. Karp & M. Pinsky, First order asymptotics of the principal eigenvalue of a tubular neighborhood, *Contemp. Math.*, Vol **73** (1988), 105-120.

Parallel transport of Brownian motion

[17] M. Liao & M. Pinsky, Stochastic parallel transport for Riemannian Brownian motion, *Contemp. Math.*, Vol **73** (1988), 203-211.

Volume of a small ball

[18] A. Gray & L. Van Hecke, Riemannian geometry as determined by the volumes of small geodesic balls, *Acta Math.*, **142** (1979), 157-198.

[19] L. Karp & M. Pinsky, The volume of a small extrinsic ball in a submanifold, *Bull. London Math. Soc.*, **21** (1989), 87-92.

Expository works

[20] M. Pinsky, Stochastic Riemannian Geometry, in *Probabilistic Analysis*, Vol **1**, Academic Press (1978), 199-236.

[21] M. Pinsky, Can you feel the shape of a manifold with Brownian motion, *Exposition. Math.*, **2** (1984), 263-271.

[22] M. Pinsky, Mean exit times and hitting probabilities for small geodesic balls,

Proceedings of the First BiBos Symposium, Springer Verlag Lecture Notes in Mathematics, Vol **1158** (1985), 216-223.

[23] M. Pinsky, Local stochastic differential geometry, *Contemp. Math.*, Vol **73** (1988), 263-272.

Books on differential geometry

[24] M. Berger, P. Gauduchon & E. Mazet, *Spectre d'une varieté Riemannienne*, Springer Verlag Lecture Notes in Mathematics, Vol **194**, 1971.

[25] J. Cheeger & D. Ebin, *Comparison theorems in Riemannian geometry*, North Holland, 1975.

[26] S. Kobayashi & K. Nomizu, *Foundations of differential geometry* (2 Vols), Interscience, 1963.

[27] A. Gray, *Tubes*, Addison Wesley, 1990.

Estimates on the heat kernel for second order divergence form operators

Daniel W. Stroock

Lecture I : Some basic facts

Throughout this lecture, N is a fixed positive integer, λ is a fixed number in the interval $(0,1]$, and $a : \mathbb{R}^N \longrightarrow \mathbb{R}^N \otimes \mathbb{R}^N$ will denote a generic, smooth, symmetric matrix valued function which satisfies

$$\lambda I_{\mathbb{R}^N} \le a(x) \le \frac{1}{\lambda} I_{\mathbb{R}^N}, \qquad x \in \mathbb{R}^N \tag{I.1}$$

in the sense of non-negative matrices. Associated with such an a, one can consider two elliptic differential operators:

$$a\nabla^2 = \sum_{i,j=1}^{N} a^{i,j}(x) \frac{\partial^2}{\partial x^i \partial x^j} \tag{I.2}$$

and

$$\nabla \cdot (a\nabla) = \sum_{i,j=1}^{N} \frac{\partial}{\partial x^i}\Big(a^{i,j}(x)\frac{\partial}{\partial x^j}\Big); \tag{I.3}$$

and each of these operators determines a unique Markov semigroup on $C_b(\mathbb{R}^n;\mathbb{R})$. Namely, if L denotes either $a\nabla^2$ or $\nabla \cdot (a\nabla)$, then there is a unique semigroup of bounded operators $\{P_t^L : t > 0\}$ on $C_b(\mathbb{R}^n;\mathbb{R})$ such that

$$[P_t^L \phi](x) = \phi(x) + \int_0^t [P_s^L L\phi](x)\, ds, \quad t > 0 \text{ and } \phi \in C_b^2(\mathbb{R}^N;\mathbb{R}). \tag{I.4}$$

In fact, by the weak maximum principle, each operator maps non-negative functions into non-negative functions and takes constant functions into themselves. Furthermore, if $\mathcal{S}(\mathbb{R}^N;\mathbb{R})$ denotes the Schwartz test-function space of smooth functions all of whose derivatives are rapidly decreasing, then each P_t^L maps $\mathcal{S}(\mathbb{R}^N;\mathbb{R})$ into itself and

$$(t,\phi) \in (0,\infty) \times \mathcal{S}(\mathbb{R}^N;\mathbb{R}) \longmapsto [P_t^L \phi] \in \mathcal{S}(\mathbb{R}^N;\mathbb{R})$$

is continuous. In particular, by (I.4), we have both

$$[P_{t+h}^L \phi](x) - [P_t^L \phi](x) = \int_0^h [P_{t+s}^L \circ L\phi](x)\, ds$$

Probability Theory

and

$$[P^L_{t+h}\phi](x) - [P^L_t\phi](x) = \int_0^h [P^L_s \circ L \circ P^L_t\phi](x)\, ds$$

for all $(t,x) \in (0,\infty) \times \mathbb{R}^N$ and $\phi \in S(\mathbb{R}^N;\mathbb{R})$. Hence

$$(I.5) \qquad \frac{d}{dt}[P^L_t\phi] = [P^L_t L\phi] = [LP^L_t\phi] \qquad t > 0 \text{ and } \phi \in S(\mathbb{R}^N;\mathbb{R}).$$

Finally, by the Riesz Representation Theorem, we know that there is a unique transition probability function $P^L(t,x,\cdot)$ for which

$$(I.6) \qquad [P^L_t\phi](x) = \int_{\mathbb{R}^N} \phi(y)\, P^L(t,x,dy), \qquad (t,x) \in (0,\infty) \times \mathbb{R}^N$$

for every $\phi \in C_b(\mathbb{R}^N;\mathbb{R})$.

In these lectures we will be reporting on some of the properties of the semigroups $\{P^L_t : t > 0\}$. Our emphasis will be on those properties which are expressed most clearly in terms of the transition probability functions $P^L(t,x,\cdot)$ and are therefore of most immediate interest to probabilists. Because we will be restricting our attention to properties which depend only on the constant λ and not on the quality (i.e., the smoothness) of the function a, it will be essential for us to base our reasoning on basic structural facts about the elliptic operators $a\nabla^2$ and $\nabla \cdot (a\nabla)$. As a preliminary, we begin with the following lemma which, in spite of its simplicity, gives a useful introduction to the type of reasoning involved.

Notation. Throughout, $\|\psi\|_p$ will be used to denote the $L^p(\mathbb{R}^N)$-norm of a function ψ and (ϕ,ψ) will be used for the $L^2(\mathbb{R}^N)$-inner product of ϕ with ψ. Similarly, when Φ and Ψ are $\mathbb{R}^N$-valued functions, we will write (Φ,Ψ) for their inner product in $L^2(\mathbb{R}^N;\mathbb{R}^N)$. Finally, if T is a linear operator, $\|T\|_{p\to q}$ will denote the norm of an operator T as map from $L^p(\mathbb{R}^N)$ into $L^q(\mathbb{R}^N)$.

Lemma 1.7. *For $\theta \in \mathbb{R}^N$ set $e_\theta(x) = \exp[\theta \cdot x]$. Then, for all $t > 0$ and $\phi \in S(\mathbb{R}^N;\mathbb{R}^N)$*

$$(I.8) \qquad \|e_{-\theta}P^L_t(e_\theta\phi)\|_\infty \le \exp[t|\theta|^2/\lambda]\|\phi\|_\infty \quad \textit{if } L = a\nabla^2$$

and

$$(I.9) \qquad \|e_{-\theta}P^L_t(e_\theta\phi)\|_2 \le \exp[t|\theta|^2/\lambda]\|\phi\|_2 \quad \textit{if } L = \nabla\cdot(a\nabla).$$

Proof. Observe that (I.8) comes down to showing that

$$e_{-\theta}(x)[P^L_t e_\theta](x) \le \exp[t|\theta|^2/\lambda],$$

and, because our hypotheses are invariant under translation, this need only be done when $x = 0$. Next, let $t > 0$ be given, and consider the function

$$u(s) = \exp[(t-s)|\theta|^2/\lambda][P_s^L e_\theta](0) \quad \text{for} \quad s \in [0,t].$$

Clearly the desired inequality is just the statement that $u(t) \leq u(0)$, and by (I.5), we see that $u'(s) \leq 0$.

In proving (I.9), let $\phi \in S(\mathbb{R}^N;\mathbb{R})$ be given and set $\phi_t^\theta = e_{-\theta}[P_t^L(e_\theta \phi)]$. Then, again by (I.5),

$$\begin{aligned}
\frac{d}{dt}\|\phi_t^\theta\|_2^2 &= 2(e_{-\theta}\phi_t^\theta, \nabla\cdot(a\nabla)\phi_t^\theta)) \\
&= -2(\nabla(e_{-\theta}\phi_t^\theta), a\nabla(e_\theta\phi_t^\theta)) \\
&= -2(\nabla\phi_t^\theta, a\nabla\phi_t^\theta) + 2\int_{\mathbb{R}^N}(\theta\cdot a\theta)(\phi_t^\theta)^2\, dx \leq \frac{2|\theta|^2}{\lambda}\|\phi_t^\theta\|_2^2,
\end{aligned}$$

and clearly the estimate follows immediately from this. □

Comments. (i) As the proof of Lemma I.7 makes clear, the structural advantage possessed by the operators $a\nabla^2$ is that one can control the size of $L\phi$ in such a way that the estimate depends on a only through λ; and from this type of control one easily gets estimates on $\|P_t^L\|_{\infty\to\infty}$. On the other hand, when $L = \nabla\cdot(a\nabla)$, the control which one has is quite different. Indeed, now the size of $L\phi$ certainly cannot be controlled in terms of λ alone but, instead, it is the quadratic form $(\phi, -L\phi)$ which is easy to control. As a consequence of this difference, it is $\|P_t^L\|_{2\to 2}$, and not $\|P_t^L\|_{\infty\to\infty}$, which is easy to estimate when $L = \nabla\cdot(a\nabla)$.

(ii) Before moving on, it may be helpful to interpret the estimates (I.8) and (I.9) in terms of $P^L(t,x,\cdot)$. The interpretation of (I.8) is obvious. Namely, it says that

$$\int_{\mathbb{R}} e^{\theta\cdot(y-x)} P^L(t,x,dy) \leq e^{t|\theta|^2/\lambda}, \qquad (t,x)\in(0,\infty)\times\mathbb{R}^N \text{ and } \theta\in\mathbb{R}^N;$$

and, after an elementary application of Chebyshev's inequality, this leads to

$$P^L(t,x,B(x,r)^c) \leq 2N\exp\left[-\frac{r^2}{4N\lambda t}\right], \qquad r>0,$$

for all $(t,x)\in(0,\infty)\times\mathbb{R}^N$. (Here, and elsewhere, $B(x,r)$ denotes the ball in $\mathbb{R}^N$ of radius r around the point x.) The interpretation of (I.9) in terms of $P^L(t,x,\cdot)$ is not as satisfactory. In fact, essentially the only thing that one can do is re-write it in the equivalent form

$$\iint_{\mathbb{R}^N\times\mathbb{R}^N} \phi(x)\psi(y)e^{\theta\cdot(y-x)}\, P^L(t,x,dy)\, dx \;\leq\; \exp[t|\theta|^2/\lambda]\; \|\phi\|_2\|\psi\|_2$$

for all $t > 0$ and $\theta \in \mathbb{R}^N$.

(iii) One final remark about Lemma I.7 will be important. Neither (I.8) nor (I.9) says anything about the possible *smoothing* properties of $\{P_t^L : t > 0\}$. That is, neither one says that the action of P_t^L does anything to *improve* the quality of the functions; they are both statements which only say that it does functions no particular harm. Of course, a look at the proof of Lemma I.7 quickly reveals why one should not expect to get smoothing estimates from the sort of considerations there: namely, we used only the right hand side of (I.1) and ignored the left hand side entirely. Learning how to use the lower bound on the left hand side of (I.1) is the key to what analysts call *elliptic regularity theory*, and the reasoning involved is often quite subtle. In fact, when $L = a\nabla^2$, it is not at all obvious how one can bring the lower bound into consideration. On the other hand, when $L = \nabla \cdot (a\nabla)$, it is quite clear where the lower bound could have been used effectively. Indeed, in our derivation of (I.9), we simply said that the term

$$-2(\nabla\phi_t^\theta, a\nabla\phi_t^\theta)$$

is non-positive; whereas, the left hand side of (I.1) gives us a strong quantitative way to measure just how negative this term actually is.

We will close this lecture with a summary of a few estimates which $P^L(t,x,\cdot)$ is known to satisfy. For purposes of comparison, we state these results for both $L = a\nabla^2$ and $L = \nabla \cdot (a\nabla)$; however, in the ensuing lectures we will restrict our attention to the easier case $L = \nabla \cdot (a\nabla)$.

Theorem I.10. *There is a (unique) $p^L \in C^\infty((0,\infty) \times \mathbb{R}^N \times \mathbb{R}^N; (0,\infty))$ for which $P^L(t,x,dy) = p^L(t,x,y)\,dy$; and, when $L = \nabla\cdot(a\nabla)$, $p^L(t,x,y) = p^L(t,y,x)$. Furthermore, for each $\lambda \in (0,1)$ there is a $q(\lambda,N) \in [N',\infty]$ (throughout, q' denotes the Hölder conjugate of q) and a map*

$$q \in (1, q(\lambda,N)) \longmapsto M_q \in [1,\infty)$$

such that

$$(I.11) \qquad \|p^{a\nabla^2}(t,x,\cdot)\|_q \le M_q t^{-N/2q'} \quad (t,x) \in (0,\infty) \times \mathbb{R}^N$$

for each $q \in [1, q(\lambda,N)]$. On the other hand, for each $q > N'$ there is a $\lambda \in (0,1)$ for which

$$(I.12) \qquad \sup_a \|p^{a\nabla^2}(1,0,\cdot)\|_q = \infty,$$

where the supremum is taken over smooth a's satisfying (I.1). Finally, for each $\lambda \in (0,1)$ there is an $M = M(\lambda,N) \in [1,\infty)$ such that

$$(I.13) \qquad \frac{1}{Mt^{N/2}} e^{-M|y-x|^2/t} \le p^{\nabla\cdot(a\nabla)}(t,x,y) \le \frac{M}{t^{N/2}} e^{-|y-x|^2/Mt},$$

for all $(t,x,y) \in (0,\infty) \times \mathbb{R}^N \times \mathbb{R}^N$.

The results contained in Theorem I.10 are the culmination of approximately thirty years of work on the subject. As was mentioned above, the case when $L = \nabla \cdot (a\nabla)$ is easier and so it is not surprising that it is also the one for which the history goes back furthest. In fact, the first article in which estimates which depend only on λ were obtained for $p^{\nabla \cdot (a\nabla)}(t,x,y)$ was J. Nash's famous 1958 paper. Shortly thereafter, first deGiorgi and then J. Moser advanced Nash's program and brought it to the state where (except for minor improvements) it is today. In particular, it was Moser who proved a quantitative version of Harnack principle for the operator $\frac{\partial}{\partial t} - \nabla \cdot (a\nabla)$, and it was Moser's result on which D. Aronson based his proof of the lower bound in (I.13). The evolution of (I.11) and (I.12) was much later and slower. Among the first articles to appear on such matters were those of A.D. Aleksandrov and C. Pucci; but it was not until the appearance of the groudbreaking articles of N. Krylov and M.V. Safonov that estimates approaching (I.13) started to emerge. In fact, although (I.12) is implicit in Pucci's article, both it and (I.11) appears for the first time in [F & S, 1]. For precise references to the work cited above, the interested reader should consult the bibliography to [G & T], which is itself an excellent reference for this sort of analysis. Finally, for those who are particularly interested in the operator $\nabla \cdot (a\nabla)$, there is an excellent new book [D, 2] by E.B. Davies.

Lecture II : The upper bound

As I said at the end of the first lecture, we will be concentrating our attention on the case when $L = \nabla \cdot (a\nabla)$. Thus, throughout this lecture, $L = \nabla \cdot (a\nabla)$ and $p^L(t,x,y)$ is therefore a smooth, symmetric, positive, transition probability kernel.

The goal of this lecture is to describe a way to prove the upper bound in (I.13). As we pointed out in comment (iii) following Lemma I.7, a great deal was lost in our derivation of (I.9) when we estimated $-2(\nabla\phi_t^\theta, a\nabla\phi_t^\theta)$ from above by 0. In order to improve our estimate, we will use the following elementary form of a Sobolev inequality.

Lemma II.1. (Nash) *There is a* $C_N \in (0,\infty)$ *with the property that*

$$\|\phi\|_2^{2+4/N} \le C_N \|\nabla\phi\|_2^2 \, \|\phi\|_1^{4/N}, \quad \phi \in S(\mathbb{R}^N;\mathbb{R}). \tag{II.2}$$

Proof. Set

$$g_t(x) = (4\pi t)^{-N/2} \exp\left[-\frac{|x|^2}{4t}\right], \quad (t,x) \in (0,\infty) \times \mathbb{R}^N. \tag{II.3}$$

Next, let $\phi \in S(\mathbb{R}^N;\mathbb{R})$ be given, and set $\phi_t = g_t * \phi$ (the "$*$" denotes convolution). Then, $\|\phi_t\|_1 \le \|\phi\|_1$, and, because all the operations involved are translation invariant and therefore commute with one another,

$$(\phi, \Delta\phi_t) = -(g_{t/2} * (\nabla\phi), g_{t/2} * (\nabla\phi)) \ge -\|\nabla\phi\|_2^2.$$

Thus, from $\phi_t = \phi + \int_0^t \Delta\phi_s \, ds$, we see that

$$\frac{\|\phi\|_1^2}{(4\pi t)^{N/2}} \ge (\phi,\phi_t) = \|\phi\|_2^2 + \int_0^t (\phi, \Delta\phi_s)\, ds \ge \|\phi\|_2^2 - t\|\nabla\phi\|_2^2$$

and therefore

$$\|\phi\|_2^2 \le (4\pi t)^{-N/2}\|\phi\|_1^2 + t\|\nabla\phi\|_2^2, \quad t \in [0,\infty).$$

Now make the optimal choice of t in the preceding, and thereby arrive at (II.2). □

Remark. It is worth noting that the preceding derivation of (II.2) is no more than a trivial application of the upper bound in (I.13) for the special case when $a \equiv I_{\mathbb{R}^N}$. What makes this observation interesting is that, as we about to see, (II.2) essentially implies the right hand side of (I.13) for arbitrary a satisfying (I.1); and therefore, in a sense, what we are showing is that the upper bound for $a \equiv I_{\mathbb{R}^N}$ implies the upper bound for all a's.

In order to develop an appreciation for the power of the estimate (II.2), it may be helpful to see that it immediately implies the estimate $p^L(t,x,y) \le M/t^{N/2}$.

To this end, let $\phi \in S(\mathbb{R}^N;\mathbb{R}) \setminus \{0\}$ and set $\phi_t = [P_t^L \phi]$. Then, by Minkowski's inequality,

$$\|\phi_t\|_q = \left(\int_{\mathbb{R}^N} \left|\int_{\mathbb{R}^N} p^L(t,x,y)\phi(y)\,dy\right|^q dx\right)^{1/q} \le \|\phi\|_q, \quad q \in [1,\infty].$$

In addition, just as in the proof of Lemma I.7,

$$\frac{d}{dt}\|\phi_t\|_2^2 \le -2\lambda\|\nabla\phi_t\|_2^2,$$

which, together with (II.2) and $\|\phi_t\|_1 \le \|\phi\|_1$, means that

$$\frac{d}{dt}\|\phi_t\|_2 \le -\frac{\lambda}{C_N\,\|\phi\|_1^{4/N}}\,\|\phi_t\|_2^{1+4/N}.$$

Hence, after solving this easy differential inequality, we see that

$$\|\phi_t\|_2 \le \left(\frac{NC_N}{4\lambda t}\right)^{N/4}\|\phi\|_1,$$

which, when expressed in terms of the kernel p^L, says that

$$\sup_{y\in\mathbb{R}^N}\|p^L(t,\cdot,y)\|_2 \le \left(\frac{NC_N}{4\lambda t}\right)^{N/4}.$$

But, by symmetry and the Chapman–Kolmogorov equation, this immediately implies that

$$\begin{aligned} p^L(t,x,y) &= \int_{\mathbb{R}^N} p^L(t/2,x,\xi)p^L(t/2,\xi,y)\,d\xi \\ &\le \|p^L(t/2,x,\cdot)\|_2\,\|p^L(t/2,\cdot,y)\|_2 \le \left(\frac{NC_N}{2\lambda t}\right)^{N/2}. \end{aligned}$$

At first, one might hope that the trivial argument carried out at the end of the preceding remark might apply equally well to (cf. the proof of Lemma I.7) ϕ_t^θ and lead first to

$$\|e_{-\theta}p^L(t,\cdot,y)e_\theta(y)\|_2 \le \left(\frac{NC_N}{4\lambda t}\right)^{N/4} e^{t|\theta|^2/\lambda},$$

thence to

$$e^{-\theta\cdot x}p^L(t,x,y)e^{\theta\cdot y} \le \left(\frac{NC_N}{2\lambda t}\right)^{N/2} e^{t|\theta|^2/\lambda},$$

and finally, after one moves the exponentials to the right side and then makes the optimal choice of θ, to

$$p^L(t,x,y) \le \left(\frac{NC_N}{2\lambda t}\right)^{N/2}\exp\left[-\frac{|y-x|^2}{4\lambda t}\right].$$

However, a moment's reflection reveals a serious flaw right at the beginning of the preceding line of reasoning. Namely, except when $\theta = 0$, we no longer know that $\|\phi_t^\theta\|_1 \le \|\phi\|_1$. For this reason, we must adopt a slightly different approach; and, so long as we are at it, we will do some fine tuning of the basic argument used in the derivation of (I.9).

Lemma II.4. *Given a* $\psi \in C_b^\infty(\mathbb{R}^N;\mathbb{R})$, *set* $e_\psi(x) = e^{\psi(x)}$, *and, for non-negative* $\phi \in S(\mathbb{R}^N;\mathbb{R}) \setminus \{0\}$, *define*

$$\phi_t^\psi = e_{-\psi}[P_t^L(e_\psi \phi)], \quad t > 0.$$

Then, for each $q \in [1,\infty)$,

$$\frac{d}{dt}\|\phi_t^\psi\|_{2q} \le -\frac{\lambda}{qC_N}\frac{\|\phi_t^\psi\|_{2q}^{1+4q/N}}{\|\phi_t^\psi\|_q^{4q/N}} + q\Gamma_a(\psi)^2\|\phi_t^\psi\|_{2q}, \tag{II.5}$$

where

$$\Gamma_a(\psi) \equiv \sup_{x\in\mathbb{R}^N}(\nabla\psi(x)\cdot a(x)\nabla\psi(x))^{1/2}.$$

Proof. There is nothing new involved in the derivation of (II.5), it is just a somewhat more tedious re-run of the computation used in the proof of (I.9) combined with (II.2) applied to the function $(\phi_t^\psi)^q$.

Continuing with the notation in Lemma II.4, set

$$u_n(t) = \|\phi_t^\psi\|_{2^n} \quad \text{for } n \in \mathbb{N} \text{ and } t \ge 0.$$

Starting from (II.5), an elementary computation allows one to check that there is a $K = K(\lambda, N) \in (0,\infty)$ with the property that, for every $\delta \in (0,1]$,

$$u_{n+1}(t) \le \left(\frac{4^n K}{\delta t}\right)^{\frac{N}{4}\frac{1}{2^n}} \exp\left[\frac{t\delta\Gamma_a(\psi)^2}{2^n}\right] u_n(t), \quad n \in \mathbb{N}.$$

Because, as is obvious from (II.5) with $q = 1$,

$$u_0(t) \le e^{t\Gamma_a(\psi)^2}\|\phi_2\|_2,$$

one finds that, for some other $K = K(\lambda, N) \in (0,\infty)$,

$$u_{n+1}(t) \le \frac{K}{(\delta t)^{\nu_n N/4}}\exp[t(1+\delta\nu_n)\Gamma_a(\psi)^2]\|\phi\|_2 \quad \text{with } \nu_n = 1 - \frac{1}{2^n};$$

and therefore

$$\|\phi_t^\psi\|_\infty \le \varliminf_{n\to\infty}\|\phi_t^\psi\|_{2^n} \le \frac{K}{(\delta t)^{N/4}}\exp\left[t(1+\delta)\Gamma_a(\psi)^2\right]\|\phi\|_2.$$

At this point we are essentially home free. Namely, the preceding is equivalent to

$$\sup_{x\in \mathbb{R}^N} \|e_{-\psi}(x)p^L(t,x,\cdot)e_\psi\|_2 \le \frac{K}{(\delta t)^{N/4}} \exp\left[t(1+\delta)\Gamma_a(\psi)^2\right],$$

and, after replacing ψ by $-\psi$ and using symmetry together with the Chapman-Kolmogorov equation, one concludes from this that (with a slightly different K)

$$p^L(t,x,y) \le \frac{K}{(\delta t)^{N/2}} \exp\left[\psi(x) - \psi(y) + t(1+\delta)\Gamma_a(\psi)^2\right]$$

for every $\delta \in (0,1]$ and $\psi \in C^1(\mathbb{R}^N;\mathbb{R})$; and, since $\Gamma_a(\alpha\psi) = |\alpha|\Gamma_a(\psi)$, $\alpha \in \mathbb{R}$, we can re-write the preceding as

$$(II.6) \qquad p^L(t,x,y) \le \frac{K}{(\delta t)^{N/2}} \exp\left[-\frac{d_{a^{-1}}(x,y)^2}{4(1+\delta)t}\right],$$

where

$$(II.7) \qquad d_{a^{-1}}(x,y) \equiv \sup\left\{\frac{|\psi(x)-\psi(y)|}{\Gamma_a(\psi)} : \psi \in C^1(\mathbb{R}^N;\mathbb{R})\right\}.$$

The upper bound in (I.13) results from (II.6) by considering ψ's of the form $\theta\cdot x$. However, (II.6) is a much more refined estimate than the one in (I.13). Indeed, it is not very difficult to identify $d_{a^{-1}}(x,y)$ as the *geodesic distance* between x and y relative to the Riemannian metric a^{-1}. That is,

$$(II.8) \qquad \begin{aligned} d_{a^{-1}}(x,y)^2 = \inf\Bigg\{ \int_0^1 & \dot\gamma(t)\cdot a(\gamma(t))^{-1}\dot\gamma(t)\,dt \\ & : \gamma \in C^1([0,1];\mathbb{R}^N \text{ with } \gamma(0) = x \text{ and } \gamma(1) = y\Bigg\}; \end{aligned}$$

and, as we will see in the next lecture, this means that (II.6) is quite sharp.

Comments. The derivation, given in the remark following Lemma II.1, of the estimate $p^L(t,x,y) \le C/t^{N/2}$ is essentially the same as the one given originally by Nash. Once one has this estimate, there are several ways in which one can proceed to the right hand side of (I.13). The elegant approach which we have adopted here is an adaptation of an argument introduced by E. B. Davies; and, so far as I know, it is the only one which leads to (II.6) with constants depending only on λ and N.

Lecture III : The lower bound

The main concern in this lecture will be the proof of the lower bound in (I.12). Thus, unless further notice is given, L is once again the operator $\nabla \cdot (a\nabla)$ with an a which satisfies (I.1).

It may be helpful to begin the derivation with the observation that the lower bound *ought* to be a *trivial* consequence of the upper bound. To understand this, note that, by the Chapman-Kolmogorov equation, symmetry, and Schwartz's inequality:

$$\begin{aligned} p^L(t,x,x) &= \int_{\mathbb{R}^N} p^L(t/2,x,\xi)^2\,d\xi \ge \int_{B(x,rt^{1/2})} p^L(t/2,x,\xi)^2\,d\xi \\ &\ge \frac{1}{\Omega_N r^N t^{N/2}} \left(\int_{B(x,rt^{1/2})} p^L(t/2,x,\xi)\,d\xi\right)^2 \end{aligned}$$

(we use $|S|$ to denote the Lebesgue measure of a set S and Ω_N to stand for $|B(0,1)|$); and, by the upper bound in (I.12), we can choose an $r = r(\lambda, N) \in (0,\infty)$ so that

$$\left(\int_{B(x,rt^{1/2})} p^L(t/2,x,\xi)\,d\xi\right)^2 \ge \frac{1}{2}$$

for all $(t,x) \in (0,\infty) \times \mathbb{R}^N$. Hence, there is an $\epsilon = \epsilon(\lambda, N) \in (0,\infty)$ for which

$$p^L(t,x,x) \ge \frac{\epsilon}{t^{N/2}}, \qquad (t,x) \in (0,\infty) \times \mathbb{R}^N.$$

In other words, along the diagonal the left hand side of (I.12) follows from the right hand side.

Unfortunately, getting away from the diagonal is not quite so easy. Indeed, the preceding arguement is little more than a re-statement of the fact (coming from the upper bound) that $P^L(t,x,\cdot)$ is nearly concentrated on a ball around x of radius proportional to $t^{1/2}$; but in order to extend the estimate to points (x,y) which are off the diagonal, it is necessary to know, in addition, that the mass of $P^L(t,x,\cdot)$ is spread uniformly over such a ball. Thus, what we need is to know that, by time t, the diffusion starting at x has had time to *equilibrate* on the ball $B(x,t^{1/2})$, and this will require us to have a quantitative measure of the *local ergodicity* of our diffusion. There are many ways in which one can attempt to measure ergodicity, but for our purposes the following well-known Poincaré inequality will suffice.

Lemma III.1. *Define $g_t, t > 0$, as in* (II.3). *Then*

$$\text{(III.2)} \qquad \int_{\mathbb{R}^N} g_1 \left(\psi - \int_{\mathbb{R}^N} g_1\psi\,d\xi\right)^2 d\xi \le 2\int_{\mathbb{R}^N} g_1\,|\nabla\psi|^2\,d\xi,$$

for every $\psi \in S(\mathbb{R}^N;\mathbb{R})$.

Proof. Given $\psi \in S(\mathbb{R}^N;\mathbb{R})$, set

$$\psi_t(x) = (g_{1-e^{-t}} * \psi)(e^{-t}x) \quad \text{for } (t,x) \in (0,\infty) \times \mathbb{R}^N.$$

One can then easily check that

$$\frac{d}{dt}\psi_t(x) = \left(\Delta - \frac{1}{2}x \cdot \nabla\right)\psi_t(x) = \frac{\nabla \cdot (g_1 \nabla \psi)(x)}{g_1(x)}$$

and therefore that

$$\frac{d}{dt}\int_{\mathbb{R}^N} g_1 \psi_t^2\, dx = -2\int_{\mathbb{R}^N} g_1 |\nabla \psi_t|^2\, d\xi.$$

In particular,

$$t \in [0,\infty) \longmapsto \int_{\mathbb{R}^N} g_1(\xi)[(g_{1-e^{-t}} * \psi_t)(e^{-t/2}\xi)]^2\, d\xi$$

is non-increasing for every $\phi \in S(\mathbb{R}^N;\mathbb{R})$. In fact, since

$$\nabla \psi_t(x) = e^{-t/2}(g_{1-e^{-t}} * (\nabla \psi))(e^{-t/2}x),$$

we now see that

$$\frac{d}{dt}\int_{\mathbb{R}^N} g_1 \psi_t^2\, d\xi \geq -2e^{-t}\int_{\mathbb{R}^N} g_1 |\nabla \psi|^2\, d\xi.$$

Next integrate the preceding from 0 to ∞, and, finally, note that $\psi_0 = \psi$ whereas

$$\psi_t \longrightarrow \int_{\mathbb{R}^N} g_1 \psi\, d\xi \quad \text{boundedly as } t \longrightarrow \infty. \quad \square$$

The rôle that (III.2) plays in the derivation of the lower bound is somewhat subtle. In fact, it only enters in the derivation of the following innocent looking estimate which, once again, has its origins in Nash's 1958 paper.

Lemma III.3. *There is a constant $A = A(\lambda, N) \in (0,\infty)$ with the property that*

$$\int_{\mathbb{R}^N} g_1(y) \log [p^L(t,x,x+\xi)]\, d\xi \geq -A, \quad x \in \mathbb{R}^N. \tag{III.4}$$

Proof. Without loss in generality, we will assume that $x = 0$. Set $u_t = u(t,\cdot) = p^L(t,0,\cdot)$ and

$$G(t) = \int_{\mathbb{R}^N} g_1 \log [u_t]\, d\xi.$$

We want to estimate $G(1)$ from below, and, while doing so, it may be helpful to have observed that (by Jensen's inequality)

$$G(t) \leq \log \left[\int_{\mathbb{R}^N} g_1(\xi) u(t,\xi)\, d\xi \right] \leq 0.$$

Next, after a few elementary integrations by parts followed by elementary manipulation of quadratics, note that

$$\begin{aligned} G'(t) &= (g_1 \nabla(\log u_t), a\nabla(\log u_t)) - (g_1 \nabla(\log g_1), a\nabla(\log u_t)) \\ &\geq \frac{\lambda}{2} \int_{\mathbb{R}^N} g_1 |\nabla(\log u_t)|^2\, d\xi - \frac{1}{4\lambda} \int_{\mathbb{R}^N} |\xi|^2 g_1(\xi)\, d\xi \\ &\geq \lambda \int_{\mathbb{R}^N} g_1 (\log u_t - G(t))^2\, d\xi - \frac{1}{2\lambda}, \end{aligned}$$

where, in the passage to the last line, we have used (III.2). At this point we need to make sure that the $G(t)$ in the preceding does not get *swamped* by the $\log\ u_t$. To this end, we first use the upper bound in (I.12) to choose an $r = r(\lambda, N) \in (0,\infty)$ with the property that

$$\int_{|\xi|\geq r} p^L(t,0,\xi)\, d\xi \leq \frac{1}{4} \quad \text{for} \quad t \in (0,1].$$

Then, again by the upper bound in (I.12), for any $K \in (0,\infty)$ and $t \in [1/2, 1]$,

$$\frac{3}{4} \leq \int_{B(0,r)} u(t,\xi)\, d\xi \leq \Omega_N r^N e^{-K} + 2^N M |\{\xi \in B(0,r) : u(t,\xi) \geq e^{-K}\}|,$$

Hence, we can choose $K = K(\lambda, N) \in (0,\infty)$ so that

$$|\{\xi \in B(0,r) : u(t,\xi) \geq e^{-K}\}| \geq \epsilon, \quad t \in [1/2, 1]$$

for some strictly positive $\epsilon = \epsilon(\lambda, N)$. After combining this with the lower bound which we already have for $G'(t)$, we conclude that there exists a $B = B(\lambda, N) \in [1,\infty)$ for which

$$G'(t) \geq -B^2 + \frac{G(t)^2}{B^2} \quad \text{for } t \in [1/2, 1];$$

and from here (III.4) is an elementary exercise in ordinary differential inequalities. □

As unimpressive as the estimate in (III.4) may look, it turns out to be the essential ingredient in the derivation of the lower bound. However, before we put it to work we will make one more preparation.

Let ρ be a non-negative element of $S(\mathbb{R}^N; \mathbb{R})$ which satisfies

$$(III.5) \qquad \int_{\mathbb{R}^N} \rho\, d\xi = 1 \quad \text{and} \quad H(\rho) \equiv \int_{\mathbb{R}^N} \frac{|\nabla\rho|^2}{\rho}\, d\xi < \infty.$$

Next, given a path $\gamma \in C^1([0,t]; \mathbb{R}^N)$ for some $t \in (0,\infty)$, define

$$E_{a,\rho}(\gamma) = \int_0^t \dot\gamma(s) \cdot (\rho * a^{-1}(\gamma(s)))\dot\gamma(s)\, ds.$$

Lemma III.6. *For any $(t,x) \in (0,\infty) \times \mathbb{R}^N$ and $\gamma \in C^1([0,t]; \mathbb{R}^N)$ with $\gamma(0) = x$, one has that*

$$\int_{\mathbb{R}^N} \rho(\xi) \log \left[\frac{p^L(t+\tau, x, \xi + \gamma(t))}{p^L(\tau, x, \xi + x)}\right] d\xi \geq -\frac{H(\rho)t}{3\lambda\delta} - \frac{1+\delta}{4} E_{a,\rho}(\gamma)$$

for every $\delta \in (0,1]$ and $\tau \in (0,\infty)$.

Proof. Once again, we will assume that $x = 0$. Clearly it suffices to show that

$$\frac{d}{dt}\int_{\mathbb{R}^N} \rho(\xi) \log [p^L(t+\tau, 0, \xi + \gamma(t))]\, d\xi$$

$$\geq -\frac{H(\rho)}{2\lambda\delta} - \frac{(1+\delta)}{4}\dot\gamma(t) \cdot (\rho * a^{-1}(\gamma(t)))\dot\gamma(t). \tag{III.7}$$

In proving (III.7), set $u_t = u(t,\cdot) \equiv p^L(t, 0, \cdot + \gamma(t))$ and $a_t = a(\cdot + \gamma(t))$. Then, after integrating by parts and completing squares, we find that

$$\begin{aligned}
&\frac{d}{dt}\int_{\mathbb{R}^N} \rho \log u_t\, d\xi \\
&= -\left(\nabla\left(\frac{\rho}{u_t}\right), a_t \nabla u_t\right) - (\rho\dot\gamma(t), \nabla(\log u_t)) \\
&= (\rho\nabla(\log u_t), a_t\nabla(\log u_t)) - (\rho\nabla(\log u_t), a_t(\nabla(\log\rho) + a_t^{-1}\dot\gamma(t))) \\
&\geq -\frac{1}{4}(\rho(\nabla(\log\rho) + a_t^{-1}\dot\gamma(t)), a_t(\nabla(\log\rho) + a_t^{-1}\dot\gamma(t)));
\end{aligned}$$

and, from here, (III.7) is easy. □

With these preliminaries, we can now prove the lower bound in (I.12), and more! To complete (I.12), take $\rho = g_1$, $\delta = \tau = 1$, and $\gamma(t) = x + (y-x)t$ in Lemma III.6. Then, by (III.4) and the fact that $H(g_1) = 1/2$, we see that

$$\int_{\mathbb{R}^N} g_1(\xi) \log \left[p^L\left(\frac{3}{2}, x, \xi + \frac{x+y}{2}\right)\right] d\xi \geq -A - \frac{1}{4\lambda} - \frac{|y-x|^2}{4\lambda}.$$

Similarly, with the same choice of ρ, δ, and τ but with $\gamma(t) = y - (y-x)t$, we also have that

$$\int_{\mathbb{R}^N} g_1(\xi) \log \left[p^L\left(\frac{3}{2}, y, \xi + \frac{x+y}{2}\right)\right] d\xi \geq -A - \frac{1}{4\lambda} - \frac{|y-x|^2}{4\lambda}.$$

Hence, by the Chapman–Kolmogorov equation, translation invariance of Lebesgue's measure, symmetry, and Jensen's inequality:

$$\begin{aligned}
&\log\left[\frac{p^L(3,x,y)}{(4\pi)^{N/2}}\right]\\
&\quad = \log\left[\frac{1}{(4\pi)^{N/2}}\int_{\mathbb{R}^N} p^L\left(\frac{3}{2},x,\xi+\frac{x+y}{2}\right)p^L\left(\frac{3}{2},y,\xi+\frac{x+y}{2}\right)d\xi\right]\\
&\quad \geq \log\left[\int_{\mathbb{R}^N} g_1(\xi)p^L\left(\frac{3}{2},x,\xi+\frac{x+y}{2}\right)p^L\left(\frac{3}{2},y,\xi+\frac{x+y}{2}\right)d\xi\right]\\
&\quad \geq -2A-\frac{1}{2\lambda}-\frac{|y-x|^2}{2\lambda};
\end{aligned}$$

and clearly this means that there is an $\epsilon=\epsilon(\lambda,N)\in(0,\infty)$ for which

$$p^L(3,x,y)\geq\epsilon\exp\left[-\frac{|y-x|^2}{2\lambda}\right].$$

To complete the derivation, observe that (by uniqueness), one can easily check that, for any $\alpha\in(0,\infty)$,

$$p^L(\alpha t,\alpha^{1/2}x,\alpha^{1/2}y)=\frac{1}{\alpha^{N/2}}p^{L^\alpha}(t,x,y),$$

where L^α is the operator corresponding to the coefficient matrix $a(\alpha^{1/2}\cdot)$, and, in conjunction with what we already have, this is more than enough to produce an $M=M(\lambda,N)\in(0,\infty)$ for which the left hand side of (I.12) holds.

Although we have now completed the proof of (I.12), the lower bound which we have produced is a poor companion to the upper bound in (II.6). To remedy this situation, let ρ satisfying (III.5) be given, and set

$$S(\rho)=\int_{\mathbb{R}^N}\rho(\xi)|\xi|^2\,d\xi. \tag{III.8}$$

By the lower bound which we already have,

$$\int_{\mathbb{R}^N}\rho(\xi)\log\left[p^L(t,x,x+\xi)\right]d\xi\ \geq\ \log\left[\frac{1}{Mt^{N/2}}\right]-\frac{MS(\rho)}{t};$$

and therefore, by Lemma (III.6),

$$\begin{aligned}
&\int_{\mathbb{R}^N}\rho(\xi)\log\left[p^L((1+\delta)t,x,\xi+\gamma(t))\right]d\xi\\
&\qquad\geq\log\left[\frac{1}{Mt^{N/2}}\right]-\frac{MS(\rho)}{\delta t}-\frac{H(\rho)t}{2\lambda\delta}-\frac{1+\delta}{4}E_{a,\rho}(\gamma)
\end{aligned}$$

for any $(t,x)\in(0,\infty)\times\mathbb{R}^N$, $\delta\in(0,1]$, and $\gamma\in C^1([0,t];\mathbb{R}^N)$ with $\gamma(0)=x$. Now let $(t,x,y)\in(0,\infty)\times\mathbb{R}^N\times\mathbb{R}^N$ be given, and suppose that $\gamma\in C^1([0,t];\mathbb{R}^N)$

satisfies $\gamma(0) = x$ and $\gamma(t) = y$. Then, by the preceding applied to $\gamma|_{[0,t/2]}$ and $\hat{\gamma}|_{[0,t/2]}$, with $\hat{\gamma} \equiv \gamma(t - \cdot)$:

$$\int_{\mathbb{R}^N} \rho(\xi) \log \left[p^L((1+\delta)t/2, x, \xi + \gamma(t/2))\right] d\xi$$
$$\geq \log \left[\frac{1}{Mt^{N/2}}\right] - \frac{2MS(\rho)}{\delta t} - \frac{H(\rho)t}{2\lambda\delta} - \frac{1+\delta}{4} E_{a,\rho}(\gamma|_{[0,t/2]}),$$

$$\int_{\mathbb{R}^N} \rho(\xi) \log \left[p^L((1+\delta)t/2, y, \xi + \gamma(t/2))\right] d\xi$$
$$\geq \log \left[\frac{1}{Mt^{N/2}}\right] - \frac{2MS(\rho)}{\delta t} - \frac{H(\rho)t}{2\lambda\delta} - \frac{1+\delta}{4} E_{a,\rho}(\hat{\gamma}|_{[0,t/2]}),$$

and so, by exactly the same reasoning as we used before,

$$p^L((1+\delta)t, x, y)$$
$$\geq \frac{1}{M^2 \|\rho\|_\infty t^N} \exp\left[-\frac{H(\rho)t}{\lambda\delta} - \frac{4MS(\rho)}{\delta t} - \frac{1+\delta}{4} E_{a,\rho}(\gamma)\right].$$

Notice that if $\rho_t(\xi) \equiv t^{-N/2}\rho(\xi/t^{1/2})$, then

$$H(\rho_t) = \frac{H(\rho)}{t}, \quad S(\rho_t) = tS(\rho), \quad \text{and } \|\rho_t\|_\infty = t^{-N/2}\|\rho\|_\infty.$$

Hence, the preceding leads to the existence of an $M = M(\lambda, \rho, N) \in [1, \infty)$ such that

$$p^L((1+\delta)t, x, y) \geq \frac{e^{-M/\delta}}{Mt^{N/2}} \exp\left[-\frac{1+\delta}{4} E_{a,\rho_t}(\gamma)\right], \quad \delta \in (0, 1].$$

Finally, after making the optimal choice of γ (cf. (II.8)), the above becomes the following statement.

Theorem III.9. *Let ρ be a non-negative element of $S(\mathbb{R}^N; \mathbb{R})$ which satisfies (III.5) and define $\rho_t(\xi) = t^{N/2}\rho(\xi/t^{1/2})$ for $t > 0$. Then there exists an $M \in [1, \infty)$, which depends only on λ, N, $H(\rho)$, and $S(\rho)$ (cf. (III.8)), such that, for all $(t, x, y) \in (0, \infty) \times \mathbb{R}^N \times \mathbb{R}^N$,*

$$p^L(t, x, y) \geq \frac{e^{-M/\delta}}{Mt^{N/2}} \exp\left[-\frac{(1+\delta)d_t(x, y)^2}{4t}\right], \quad \delta \in (0, 1],$$

where (cf. (II.8)) $d_t(x, y)$ is the Riemannian distance associated with the Riemannian metric $\rho_t * a^{-1}$.

Comments. Up through Lemma (III.3), the procedure adopted in this lecture is derived directly from the one used in [F & S]. However, from that point on, the argument given is quite different, and it is this difference which accounts for the sharpened form of the lower bound which is given in Theorem (III.9). The new argument was developed by J. Norris and me and will appear in a forthcoming article by the two of us.

The methodology presented in these lectures is applicable to a variety of related situations. For example, without any change in the arguement, one can see that the estimate (I.12) extends to operators $L = \omega\nabla \cdot (a\nabla)$, where $\omega : \mathbb{R}^N \longrightarrow (0,\infty)$ is a measurable function which satisfies $\lambda \leq \omega \leq \frac{1}{\lambda}$. Furthermore, with a little more work (cf. [S] and the article with Norris alluded to above), the same estimate holds, at least when $t \in (0,1]$, for the $p^L(t,x,y)$ associated with an operator L of the form

$$L = \nabla \cdot (a\nabla) + ab \cdot \nabla - \nabla(a\hat{b}) + c,$$

where $b, \hat{b} : \mathbb{R}^N \longrightarrow \mathbb{R}^N$ and $c : \mathbb{R}^N \longrightarrow \mathbb{R}$ are bounded measurable functions. For situations which are a little further removed from the classical elliptic case,the interested reader might want to have a look at a forthcoming article by me and L. Soloff–Coste in which we treat certain subelliptic operators on unimodular Lie groups having polynomial growth. Finally, an abstract treatment of the ideas in Lecture II is given in [C, K, & S].

References

[C, K, & S] E. Carlen, S. Kusuoka, and D. Stroock, Upper bounds for symmetric Markov transition functions, *Ann. Inst. H. Poincaré, Prob. Statist.* **23**(1987), 245-287.

[D, 1] E.B. Davies, Explicit constants for Gaussian upper bounds on heat kernels, *Amer. J. Math.* **109**(1987), 319-334.

[D, 2] E.B. Davies, *Heat Kernels and Spectral Theory*, Cambridge Tracts in Math. # 92, Cambridge University Press, (1989).

[F & S, 1] E.B. Fabes and D.W. Stroock, The L^p-integrability of Green's functions and fundamental solutions for elliptic and parabolic equations, *Duke Math. J.*, **51** # 4 (1984), 997-1016.

[F & S, 2] E.B. Fabes and D.W. Stroock, A new proof of Moser's parabolic Harnack inequality via the old ideas of Nash, *Arch. Rational Mech. Anal.*, **96**, (1986), 327-338.

[G & T] D. Gilbarg and N. Trudinger, *Elliptic Partial Differential Equations of Second Order*, Springer–Verlag, (1977).

[S] D.W. Stroock, Diffusion semigroups corresponding to uniformly elliptic divergence form operators, *Springer Lecture Notes in Math., Sem. de Prob.*, # XXII.

Special invited lecture

On Malliavin calculus

Kiyosi Itô

Introduction

Malliavin calculus is usually formulated on the Wiener space or more generally on an abstract Wiener space [1], [2], [3], [4]. However, the abstract Wiener space is used for the purpose of representations, and its Banach space structure has no essential meaning. The purpose of the present paper is to establish Malliavin calculus directly on the base probability space. This idea was published in our notes [5], [6], but these notes contained quite a few defects, which, we hope, will be straightened up here.

To write this note we learned a lot from S. Watanabe and I. Shigekawa, though their formulation of Malliavin calculus is based on an abstract Wiener space. We are very grateful to them for their valuable comments.

1. Preliminary concepts

Let $\Omega = (\Omega, \mathcal{F}, P)$ be a probability space.

Definition 1.1. A system of real random variables, S, is called a *Segal system* on Ω if the following three conditions are satisfied:

(S.1) Every $f \in S$ is centered Gaussian, i.e. Gauss distributed with mean 0.
(S.2) S is closed under linear combinations and almost sure limits.
(S.3) There exists a countable subsystem $C \subset S$ such that every $e \in S$ is represented in the form:

$$e = \text{a.s.} \lim_{n \to \infty} e_n \quad \text{where } \{e_1, e_2 \ldots\} \subset C.$$

From now on we fix a Segal system S on Ω. Let $\sigma[S]$ be the σ-algebra generated by S and let $L_0 = L_0(S)$ be the space of all $\sigma[S]$-measurable random variables on Ω and let

$$L_p = L_p(S) := \{f \in L_0 : E(|f|^p) < \infty\}, \quad 1 \le p < \infty.$$

Definition 1.2. A random variable $f = f(\omega)$ *on* Ω is called a *measurable function* on S if $f \in L_0(S)$, and a *p-integrable function* on S if $f \in L_p(S)$ $(1 \le p < \infty)$.

We define $< f, g > := E(fg)$ if $E(|fg|) < \infty$. Then $(L_2, <,>)$ is a separable Hilbert space and S is a closed linear subset of this space, so $(S, <,>)$ is a separable Hilbert space. If $\{e_n\}$ is orthonormal (abbr. ON) in the Hilbert space S , then $\{e_n\}$ is $i.i.d(N_{0,1})$.

Probability Theory

$(L_p, \| \|_p)(\| f \|_p = E(|f|^p)^{1/p})$ is a separable Banach space for $1 \leq p < \infty$ and $(L_0, \| \|_0)(\| f \|_0 = E(|f| \wedge 1))$ is a Fréchet space.

L_0 is a commutative ring. Also

$$L_{\infty -} = L_{\infty -}(S) = \bigcap_p L_p$$

is a subring of L_0.

Definition 1.3. $f \in L_0$ is called a *polynomial* on S if f is represented in the form:

(1) $$f = \varphi(e_1, e_2, \ldots, e_n)$$

i.e. $f(\omega) = \varphi(e_1(\omega), e_2(\omega), \ldots, e_n(\omega))$ a.s. where $\varphi \in \mathcal{P}(\mathbf{R}^n)$, i.e. φ is a polynomial on $\mathbf{R}^n$, and $e_1, e_2, \ldots, e_n \in S$.

It is easy to see that $\{e_i\}$ can be assumed to be ON in the representation (1).

We denote by $\mathcal{P} = \mathcal{P}(S)$ the set of all polynomials of S. It is obvious that

$$S \cup \mathbf{R} \subset \mathcal{P} \subset L_{\infty -} \subset L_p \subset L_1 \subset L_0.$$

$\mathcal{P}, L_{\infty -}$ and L_0 are commutative rings.

Since every $h \in S$ is Gauss distributed, we obtain

$$E(e^{\alpha |h|}) < \infty, \ \forall \alpha > 0.$$

Using this fact, we can prove the following:

Theorem 1.1. *$\mathcal{P}$ is dense in $(L_p, \| \|_p)$ for every $p \in \{0\} \cup [1, \infty)$.*

Let $\mathcal{O} = \mathcal{O}(S)$ be the *rotation group* on the Hilbert space S. Then $(Oe_1, Oe_2, \ldots, Oe_n)$ is equivalent in law to $(e_1, e_2, \ldots, e_n)$ for every $O \in \mathcal{O}$, if $e_i \in S$. Every O can be extended to a bijective map: $\mathcal{P} \to \mathcal{P}$ in the obvious way. Since $\mathcal{P}$ is dense in $(L_0, \| \|_0)$, O can be extended to a bijective map: $L_0 \to L_0$. If $f_1, f_2, \ldots, f_n \in L_0$, then $(Of_1, Of_2, \ldots, Of_n)$ is equivalent in law to $(f_1, f_2, \ldots, f_n)$. Hence

$$< Of, Og > = < f, g > \text{ if } E(|fg|) < \infty.$$

Also $O(af + bg) = aO(f) + bO(g)$, $O(fg) = Of \cdot Og$ and $\| Of \|_p = \| f \|_p$. Hence $OL_p = L_p$. Thus we see that $\mathcal{P}$ and $(L_p, \| \|_p)$ are *rotation invariant*.

2. Polynomial vector fields and polynomial 1-forms

We keep using the notations $L_p = L_p(S)$, $\| \|_p$ and $\mathcal{P} = \mathcal{P}(S)$.

Definition 2.1. A linear map $V : \mathcal{P} \to \mathcal{P}$ is called a *polynomial vector field*, written

$$V \in \mathcal{T} = \mathcal{T}(\mathcal{P})$$

if V satisfies

(V.1) (the *derivation property*) $V(fg) = V(f)g + fV(g)$
(V.2) (the *property of finite rank*) There exists a finite dimensional subspace $S_1 = S_1(V) \subset S$ *such that*

$$V(e) = 0 \text{ whenever } e \in S \cap S_1^{\perp}.$$

For $h \in S$ we define the vector field of *direction* h by

$$D_h(f) := \sum_{i=1}^{n} (\partial_i \varphi)(e_1, e_2, \dots, e_n) < e_i, h >$$

if f is represented as

$$f = \varphi(e_1, e_2, \dots, e_n), \quad e_i \in S, \varphi \in \mathcal{P}(\mathbf{R}^n). \tag{1}$$

Then $D_h : \mathcal{P} \to \mathcal{P}$ is well-defined independently of the representation (1), and it is easy to check that D_h is a polynomial vector field. It holds that

$$D_h(e) = < e, h >, \quad e \in S. \tag{2}$$

$\mathcal{T}$ is a $\mathcal{P}$-*linear space* with respect to the $\mathcal{P}$-linear combinations defined in the obvious way.

Theorem 2.1. *Every* $V \in \mathcal{T}$ *is represented in the form:*

$$V = \sum_{i=1}^{n} f_i D_{e_i}$$

where $\{e_1, e_2, \dots, e_n\}$ *is an orthonormal base (abbr. ONB) of* $S_1(V)$ *(See Definition* 2.1 (V.2)) *and* $f_i = V(e_i), i = 1, 2, \dots, n$.

Theorem 2.2. *If* $f_1, f_2, \dots, f_n \in \mathcal{P}$ *and if* $\varphi \in \mathcal{P}(\mathbf{R}^n)$, *then* $f := \varphi(f_1, f_2, \dots, f_n) \in \mathcal{P}$ *and*

$$V(f) = \sum_{i=1}^{n} (\partial_i \varphi)(f_1, f_2, \dots, f_n) V(f_i) \text{ for } V \in \mathcal{T}.$$

As a special case we obtain

$$V(f) = \sum_{i=1}^{n} (\partial_i \varphi)(e_1, e_2, \dots, e_n) V(e_i) \text{ for } V \in \mathcal{T}$$

where $f = \varphi(e_1, e_2, \dots, e_n)$, $e_i \in S$, $\varphi \in \mathcal{P}(\mathbf{R}^n)$.

Definition 2.2. A $\mathcal{P}$-linear map $\alpha \in \mathcal{T} \to \mathcal{P}$ is called a *polynomial* 1-form, written

$$\alpha \in \mathcal{P}^{1\circledast}$$

if α satisfies the *property of finite rank*, i.e. there exists a finite dimensional subspace $S_1 = S_1(\alpha) \subset S$ such that

$$\alpha(D_h) = 0 \text{ whenever } h \in S \cap S_1^{\perp}.$$

For $f \in \mathcal{P}$ we define

$$(\nabla f)(V) := V(f), \ \forall\, V \in \mathcal{T}.$$

Then ∇f is a polynomial 1-form and

$$(\nabla f)(D_h) = D_h(f), \ f \in \mathcal{P}$$

so

$$(\nabla e)(D_h) = D_h(e) = < e, h >, \ e \in S.$$

$\mathcal{P}^{1\circledast}$ is a $\mathcal{P}$-linear space with respect to the $\mathcal{P}$-linear combinations defined in the obvious way.

Theorem 2.3. *Every $\alpha \in \mathcal{P}^{1\circledast}$ is represented in the form:*

$$\alpha = \sum_{i=1}^{n} f_i \nabla e_i$$

where $\{e_1, e_2, ..., e_n\}$ is an ONB of $S_1(\alpha)$ (see Definition 2.2) and $f_i = \alpha(D_{e_i}), i = 1, 2, ..., n$.

Theorem 2.4. *If $f_1, f_2, ..., f_n \in \mathcal{P}$ and if $\varphi \in \mathcal{P}(\mathcal{R}^n)$, then $f : -\varphi(f_1, f_2, ..., f_n) \in \mathcal{P}$ and*

$$\nabla f = \sum_{i=1}^{n} (\partial_i \varphi)(f_1, f_2, ..., f_n) \nabla f_i .$$

As special cases

$$\nabla(f_1 f_2) = f_1 \nabla f_2 + f_2 \nabla f_1$$

$$\nabla(f) = \sum_{i=1}^{n} (\partial_i \varphi)(e_1, e_2, ..., e_n) \nabla e_i,$$

where $f = \varphi(e_1, e_2, ..., e_n), e_i \in S, \varphi \in \mathcal{P}(\mathbf{R}^n)$.

Definition 2.3. $\alpha \cdot \beta := \sum_{i=1}^{m} \sum_{j=1}^{n} f_i g_j < e_i, h_j >$

if $\alpha = \sum_{i=1}^{m} f_i \nabla e_i (f_i \in P, \{e_i\} : \text{ON})$ and $\beta = \sum_{j=1}^{n} g_j \nabla h_j \ (g_j \in P, \{h_j\} : \text{ON})$.

It is easy to see that $\alpha \cdot \beta$ is well-defined independently of the representations of $\alpha, \beta \in P^{1\otimes}$.

Theorem 2.5.

(i) $\alpha \cdot \beta \in P$ *for* $\alpha, \beta \in P^{1\otimes}$
(ii) $\alpha \cdot \beta$ *is* P*-bilinear in* α *and* β.
(iii) $\nabla h \cdot \nabla e = < h, e >, h, e \in S$.

Definition 2.4. For $\alpha, \beta \in P^{1\otimes}$ we define

$$< \alpha, \beta > := E(\alpha \cdot \beta).$$

It is easy to check that $(P^{1\otimes}, <,>)$ is a separable pre-Hilbert space.

Definition 2.5. For $\alpha \in P^{1\otimes}$ we define

$$|\alpha| := (\alpha \cdot \alpha)^{1/2},$$

$$\| \alpha \|_p := \| \, |\alpha| \, \|_p \, .$$

Note that $|\alpha| \in L_{\infty-}$ by $\alpha \cdot \alpha \in P$, so $\| \alpha \|_p < \infty$.

Theorem 2.6.

(i) $|\alpha \cdot \beta| \le |\alpha||\beta|, |a\alpha + b\beta| \le |a||\alpha| + |b||\beta| \quad (a, b \in R)$.
$|\alpha| = 0 \Rightarrow \alpha = 0$.

(ii) $| < \alpha, \beta > | \le \| \alpha \|_p \| \beta \|_q \quad (1 < p, q < \infty, \ p^{-1} + q^{-1} = 1)$,
$\| a\alpha + b\beta \|_p \le |a| \, \| \alpha \|_p + |b| \, \| \beta \|_q \quad (1 \le p < \infty)$,
$\| \alpha \|_p = 0 \Rightarrow \alpha = -0 \quad (p \in \{0\} \cup [1, \infty))$.

For $O \in O(S)$ we define $OV \in J$ for $V \in J$ by

$$(OV)(f) := O(v(O^{-1}f)), \ \forall f \in P.$$

Then it is easy to see that $O : T \to T$ is a linear bijection and that

$$OD_h = D_{Oh} \text{ and } O(f_1V_1 + f_2V_2) = Of_1OV_1 + Of_2OV_2.$$

Similarly we define $O\alpha$ for $\alpha \in P^{1\otimes}$ by

$$(O\alpha)(V) = O(\alpha(O^{-1}V)), \ \forall \ V \in T.$$

Then it is easy to see the following:

Theorem 2.7.

(i) $O : \mathcal{P}^{1\circledast} \to \mathcal{P}^{1\circledast}$ *is a linear bijection*
(ii) $O(\nabla f) = \nabla(Of)$,
(iii) $O(f_1\alpha_1 + f_2\alpha_2) = Of_1 O\alpha_1 + Of_2 O\alpha_2$
(iv) $O(\alpha \cdot \beta) = O\alpha \cdot O\beta$
(v) $< O\alpha, O\beta > = < \alpha, \beta >, \| O\alpha \|_p = \| \alpha \|_p$.

Thus we see that $\mathcal{T}, \mathcal{P}^{1\circledast}, <,>$ and $\| \ \|_p$ are rotation invariant.

3. The operators ∇, ∇^* and L

In the last section we defined $\nabla f \in \mathcal{P}^{1\circledast}$ for $f \in \mathcal{P}$.

Definition 3.1. The linear operator

$$\nabla : \mathcal{P} \longrightarrow \mathcal{P}^{1\circledast} \ , \ f \longmapsto \nabla f$$

is called the *gradient operator.*

It is easy to see that $\nabla f = 0 \Rightarrow f \in \mathbf{R}$.

Definition 3.2.

$$\nabla^* \alpha := -\sum_{i=1}^{n} (\nabla f_i \cdot \nabla e_i - f_i \cdot e_i)$$

if $\alpha = \sum_{i=1}^{n} f_i \nabla e_i (f_i \in \mathcal{P}, \{e_i\} :$ ON in $S)$, and the linear operator

$$\nabla^* : \mathcal{P}^{1\circledast} \longrightarrow \mathcal{P}, \alpha \longmapsto \nabla^* \alpha$$

is called the *divergent operator.*

It is easy to check that $\nabla^* \alpha$ is well-defined independently of the representation of α.

Using integration by parts, we obtain

Theorem 3.1.

$$< \alpha, \nabla f > = < \nabla^* \alpha, f > .$$

Definition 3.3. The linear operator

$$L := -\nabla^* \nabla : \mathcal{P} \longrightarrow \mathcal{P}$$

is called the *Ornstein-Uhlenbeck operator.*

Theorem 3.2.

(i) $< Lf, g >= - < \nabla f, \nabla g >=< f, Lg >$
(ii) $< Lf, f > \leq 0$ *and the equality holds only for* $f \in \mathbf{R}$.

Proof. (i) follows from the definition of L by the last theorem and (ii) follows from (i). □

From the definition we can easily deduce the following theorems.

Theorem 3.3.

$$O(\nabla f) = \nabla(Of)$$
$$O(\nabla^* \alpha) = \nabla^*(O\alpha), \alpha \in \mathcal{P}^{1\circledast}$$
$$O(Lf) = L(Of), f \in \mathcal{P}.$$

Hence ∇, ∇^* *and* L *are rotation invariant.*

Theorem 3.4. $Lf = \sum(\partial_i^2 \varphi(\underline{e}) - e_i \partial_i \varphi(\underline{e}))$
if $f = \varphi(\underline{e}), \varphi \in \mathcal{P}(\mathbf{R}^n), \underline{e} = (e_1, e_2, ..., e_n)$: ON *in* S.

4. The spectral decomposition of L

Definition 4.1. $\mathcal{H}_k := \{f \in \mathcal{P} : Lf = -kf\}, k = 0, 1, 2, ...$
is called the *Hermite subspace* of $\mathcal{P}$ of degree k.

Theorem 4.1. $\mathcal{H}_k \perp \mathcal{H}_l$ *for* $k \neq l$.

Proof. Obvious by $< Lf, g >=< f, Lg >$. □

Theorem 4.2. $\mathcal{P} = \bigoplus_{k=0}^{\infty} \mathcal{H}_k$ *(in the algebraic sense), i.e. every* $f \in \mathcal{P}$ *is uniquely represented in the form*

$$f = \sum_{k=0}^{\infty} f_k, f_k \in \mathcal{H}_k, k = 0, 1, 2, ...$$

and $f_k = 0$ *for sufficiently large* k.

Proof. Let $f = \varphi(\underline{e})$, $\underline{e} = (e_1, e_2, ..., e_n)$ is ON in S and $\varphi \in \mathcal{P}(\mathbf{R}^n)$. It suffices to check that

$$f=\sum_{i=0}^{k} f_i, f_i \in H_i, k=d(\varphi)(=\text{degree of } \varphi).$$

To prove this we use induction on k. If $k=0$, this is trivial, because if $d(\varphi)=0$, then $f \equiv c \in \mathbf{R}(=\mathcal{H}_0)$. Suppose that this is true for $d(\varphi) \leq k-1$. If $d(\varphi)=k$, then $\varphi(\underline{x})=\psi(\underline{x})+\sum_{|i|=k} a_{i_1\dots i_n} x_1^{i_1} x_2^{i_2} \dots x_n^{i_n}, \underline{x}=(x_1,\dots,x_n) \in \mathbf{R}^k$, where $d(\psi) \leq k-1$ and $|i|=i_1+i_2+\dots+i_n$. By induction $\psi(\underline{e})$ is represented in the form

$$\psi(\underline{e})=\sum_{i=0}^{k-1} g_i\ ,\ g_i \in \mathcal{H}_i.$$

Hence it suffices to observe the case where

$$\varphi(\underline{x})=x_1^{i_1} x_2^{i_2} \dots x_n^{i_n}\ \ (i_1+i_2+\dots+i_n=k),$$

so $e_\nu(\partial_\nu\varphi)(\underline{e})=i_\nu\varphi(\underline{e})$. Since

$$\begin{aligned} L(\varphi(\underline{e})) &= \Delta\varphi(\underline{e})-\sum_{\nu=1}^{n} e_\nu(\partial_\nu\varphi)(\underline{e})\ \ (\Delta\varphi:=\sum_{\nu=1}^{n}\partial^2\varphi),\\ &= \Delta\varphi(\underline{e})-\sum_{\nu=1}^{n} i_\nu\varphi(\underline{e})\\ &= \Delta\varphi(\underline{e})-k\varphi(\underline{e}), \end{aligned}$$

we obtain

$$(L+k)(\varphi(\underline{e}))=\Delta\varphi(\underline{e}).$$

Since $d(\nabla\varphi) \leq k-1$, $(L+k)(\varphi(\underline{e}))$ is represented in the form

$$(L+k)(\varphi(\underline{e}))=\sum_{i=0}^{k-1} f_i'\ ,\ f_i' \in \mathcal{H}_i$$

by induction. But

$$(L+k)f_i'=(-i+k)f_i' \text{ by } f_i' \in \mathcal{H}_i.$$

Hence

$$(L+k)(\varphi(\underline{e})-\sum_{i=0}^{k-1}\frac{1}{k-i}f_i')=(L+k)p(\underline{e})-\sum_{i=0}^{k-1} f_i'=0.$$

Let

$$f_i:=\frac{1}{k-i}f_i'\ (i=0,1,2,\dots k-1) \text{ and } f_k:=\varphi(\underline{e})-\sum_{i=0}^{k-1} f_i.$$

Then $f_i \in \mathcal{H}_i$ $(i \leq k-1)$, $(L+k)f_k=0$, so $f_k \in \mathcal{H}_k$, and $\varphi(\underline{e})=\sum_{i=0}^{k} f_i$. □

Definition 4.2. $J_k f :=$ the f_k of Theorem 4.2, $f \in \mathcal{P}, k = 0,1,2,\ldots$. J_k is the *orthogonal projection* of $\mathcal{P}$ to $\mathcal{H}_k$.

It is obvious that $J_k f = 0$ for sufficiently large k and $f = \sum_k J_k f$ for every $f \in \mathcal{P}$.

Theorem 4.3. $L = \sum_{k=0}^{\infty} (-k) J_k$,

i.e.

$$Lf = \sum_{k=0}^{\infty} (-k) J_k f \ , \ f \in \mathcal{P}$$

where $J_k f = 0$ for sufficiently large k.

Proof. This is obvious by Theorem 4.2, because $Lf = -kf$ for $f \in \mathcal{H}_k$. □

Definition 4.3. Let θ be a real-valued function on $\{0,1,2,\ldots\}$. We define $\theta(-L) : \mathcal{P} \to \mathcal{P}$ by

$$\theta(-L) := \sum_{k=0}^{\infty} \theta(-k) J_k,$$

i.e.

$$\theta(-L)(f) := \sum_{k=0}^{\infty} \theta(-k) J_k f \ , \ f \in \mathcal{P}.$$

Note that the right hand side $\sum_{k=0}^{\infty} \theta(-k) J_k f$ is actually a finite sum.

Setting $\theta(k) = e^{-kt}, (1+k)^{-t/2} (t \in \mathbf{R})$, we define

$$T_t := e^{tL} \text{ and } R_t := (I - L)^{-t/2},$$

i.e.

$$T_t := \sum_{k=0}^{\infty} e^{-kt} J_k \ , \ R_t := \sum_{k=0}^{\infty} (1+k)^{-t/2} J_k.$$

Since $J_k f = 0$ for sufficiently large $k, T_t : \mathcal{P} \to \mathcal{P}$ and $R_t : \mathcal{P} \to \mathcal{P}$ are defined for every $t \in \mathbf{R}$. Also $\{T_t\}_{t \in \mathbf{R}}$ and $\{R_t\}_{t \in \mathbf{R}}$ are one-parameter groups of operators on $\mathcal{P}$.

Theorem 4.4. $(R_t f)(\omega) = \int_0^\infty (T_s f)(\omega) \gamma_t(ds)$ a.s. for $f \in \mathcal{P}$ and $t \geq 0$ *where γ_t is a probability measure on $[0,\infty)$ defined by*

$$\gamma_t(ds) = \Gamma(t/2)^{-1} e^{-s} s^{t/2-1} ds.$$

Proof. This follows from the fact that

$$\int_0^\infty e^{-ks}\gamma_t(ds) = (1+k)^{-t/2}. \quad \square$$

5. Properties of T_t and R_t for $t \geq 0$

Let $\underline{B}_t = (B_t^1, B_t^2, ..., B_t^n)$ be an n-dimensional Wiener process on Ω independent of $\sigma[S]$. Such a process $\{\underline{B}_t, t \geq 0\}$ exists by extending Ω if necessary. Let $\underline{e} = (e_1, e_2, ..., e_n)$ be an ON system in S. Then $\underline{e}$ is N^n-distributed where $N = N_{0,1}$. Let $\underline{X}_t$ be the solution of

$$d\underline{X}_t = -\underline{X}_t dt + \sqrt{2}d\underline{B}_t\ ,\ \underline{X}_0 = \underline{e} \tag{1}$$

where this stochastic differential equation is understood with respect to the filtration

$$\mathcal{F}_t = \sigma[\underline{e}]v\sigma[\underline{B}_s, s \leq t]\ ,\ 0 \leq t < \infty.$$

Then

$$\underline{X}_t = e^{-t}\underline{e} + \int_0^t e^{-(t-s)}\sqrt{2}d\underline{B}_s. \tag{2}$$

Since $\underline{e}$ and $\{\underline{B}_t\}$ are independent, it is easy to check that

$$P(\underline{X}_t \in d\underline{x}) = N^n(d\underline{x}), \tag{3}$$

and

$$P(\underline{X}_t \in d\underline{y}|\underline{e}) = p_t^n(\underline{e}, d\underline{y}) \tag{4}$$

where $P(\cdot|\underline{e})$ is the conditional probability under $\sigma[\underline{e}]$ and

$$p_t^n(\underline{x}, d\underline{y}) = \prod_{i=1}^n N_i(dy_i), \tag{5}$$

where N_i is the Gauss measure on $\mathbf{R}$ with mean $e^t x_i$ and variance $1 - e^{-2t}$. Since

$$P(\underline{X}_t \in d\underline{y}) = E[P(\underline{X}_t \in d\underline{y}|\underline{e})],$$

we obtain

$$N^n(d\underline{y}) = \int_{\mathbf{R}^n} N^n(d\underline{x})p_t^n(\underline{x}, d\underline{y}). \tag{6}$$

Theorem 5.1. *Let* $\underline{e} = (e_1, \ldots, e_n)$ *be* ON *in* S *and* $\varphi \in \mathcal{P}(\mathbf{R}^n)$.

$$T_t(\varphi(\underline{e})) = \int_{\mathbf{R}^n} p_t^n(\underline{e}, d\underline{y})\varphi(\underline{y}) \text{ for } t \geq 0.$$

Proof. Applying the stochastic chain rule, we can derive from (1)

$$\varphi(\underline{X}_t) = \varphi(\underline{e}) + \int_0^t (L_n\varphi)(\underline{X}_s)ds + \int_0^t (\nabla_n\varphi)(\underline{X}_s)\sqrt{2}d\underline{B}_t$$

where $L_n\varphi = \sum_{i=1}^n (\partial_i^2\varphi - x_i\partial_i\varphi)$ and $\nabla_n = (\partial_1, \partial_2, \ldots, \partial_n)$. Let ψ be any bounded $\mathcal{B}^n$-measurable function on $\mathbf{R}^n$. Then

$$\begin{aligned}\psi(\underline{e})\varphi(\underline{X}_t) =& \psi(\underline{e})\varphi(\underline{e}) + \int_0^t \psi(\underline{e})(L_n\varphi)(\underline{X}_s)ds \\ &+ \int_0^t \psi(\underline{e})\nabla_n\varphi(\underline{X}_s)\sqrt{2}d\underline{B}_s.\end{aligned}$$

Hence

$$E(\psi(\underline{e})\varphi(\underline{X}_t)) = E(\psi(\underline{e})\varphi(\underline{e})) + \int_0^t E(\psi(\underline{e})(L_n\varphi)(\underline{X}_s))ds,$$

so

$$E(\varphi(\underline{X}_t)|\underline{e}) = \varphi(\underline{e}) + \int_0^t E((L_n\varphi)(\underline{X}_s)|\underline{e})ds. \tag{7}$$

Suppose that $\varphi(\underline{e}) \in \mathcal{H}_k$. Then $L(\varphi(\underline{e})) = -k\varphi(\underline{e})$, so $(L_n\varphi)(\underline{e}) = -k\varphi(\underline{e})$. This implies that $(L_n\varphi)(\underline{X}_s) = -k\varphi(\underline{X}_s)$ because $\underline{X}_s \sim_L \underline{e}$ by (3). Hence

$$E(\varphi(\underline{X}_t)|\underline{e}) = \varphi(\underline{e}) - k\int_0^t E(\varphi(X_s)|\underline{e})ds \text{ by (7)}.$$

Solving this, we obtain

$$E(\varphi(\underline{X}_t)|\underline{e}) = e^{-kt}\varphi(\underline{e}) = T_t(\varphi(\underline{e})),$$

which, combined with (4), implies

$$\int_{\mathbf{R}^n} p_t^n(\underline{e}, d\underline{y})\varphi(\underline{y}) = T_t(\varphi(\underline{e})).$$

Hence our theorem holds for $\varphi(\underline{e}) \in \mathcal{H}_k$ and so for $\varphi(\underline{e}) \in \mathcal{P}$. □

Theorem 5.2.

$$\| T_t f \|_p \leq \| f \|_p \, , \; t \geq 0 \, , \; f \in \mathcal{P} \, , \; 1 \leq p < \infty.$$

Proof. Representing $f \in \mathcal{P}$ in the form: $f = \varphi(\underline{e})$ and applying Theorem 5.1 we obtain

$$|T_t f|^p = |T_t(\varphi(\underline{e}))|^p \leq \int_{\mathbf{R}^n} p_t^n(\underline{e}, d\underline{y}) |\varphi(\underline{y})|^p$$

by Jensen's inequality.

$$\begin{aligned} \| T_t f \|_p^p &\leq \int_{\mathbf{R}^n} N^n(d\underline{x}) \int_{\mathbf{R}^n} p_t^n(\underline{x}, d\underline{y}) |\varphi(\underline{y})|^p \\ &= \int_{\mathbf{R}^n} N^n(dy) |\varphi(y)|^p \quad \text{by (6)} \\ &= \| \varphi(\underline{e}) \|_p^p = \| f \|_p^p . \quad \square \end{aligned}$$

Using this theorem and remembering that $\mathcal{P}$ is dense in $(L_p, \| \; \|_p)$, we can check that $T_t (t \geq 0)$ can be extended to a linear contraction on L_p and that the extension of T_t to L_1 coincides with that to L_p on $L_p (\subset L_1)$.

Theorem 5.3. *For every $p \in [1, \infty)$ $\{T_t, t \geq 0\}$ is a Hille-Yosida semigroup of linear contractions on L_p whose generator is the closed extension of the Ornstein-Uhlenbeck operator L on $\mathcal{P}$.*

Theorem 5.4.

$$\| R_t f \|_p \leq \| f \|_p \, , \; t \geq 0 \, , \; f \in \mathcal{P} \, , \; 1 \leq p < \infty.$$

Proof. Using Theorem 4.4 and the subadditivity of $\| \; \|_p$, we obtain

$$\| R_t f \|_p \leq \int_0^\infty \| T_s f \|_p \, \lambda_t(ds) \leq \| f \|_p \int_0^\infty \lambda_t(ds) = \| f \|_p . \quad \square$$

6. Sobolev norms on $\mathcal{P}$ and $\mathcal{P}^{1\otimes}$

Definition 6.1. Suppose that $-\infty < t < \infty$ and $1 < p < \infty$. We define the *Sobolev norm* of index (t, p) by

$$\| f \|_{t,p} := \| R_{-t} f \|_p \, , \; f \in \mathcal{P}.$$

Theorem 6.2.

$$\| f \|_{t,p} \uparrow \quad as\ t, p \uparrow .$$

Proof. If $p_1 \leq p_2$, then $\| f \|_{t,p_1} \leq \| f \|_{t,p_2}$ by the definition. If $t_1 \leq t_2$, then $t_2 - t_1 \geq 0$

$$\begin{aligned} \| f \|_{t_1,p} &= \| R_{-t_1} f \|_p = \| R_{t_2 - t_1} R_{-t_2} f \|_p \\ &\leq \| R_{-t_2} f \|_p \quad \text{(by Theorem 5.4)} \\ &= \| f \|_{t_2,p} . \qquad \square \end{aligned}$$

Theorem 6.3. *$(\mathcal{P}, \| \ \|_{t,p})$ is a normed space.*

Definition 6.2. For $\alpha = \sum_{i=1}^{n} f_i \nabla e_i, f_i \in \mathcal{P}, e_i \in S$ we define

$$\phi(-L)\alpha := \sum_{i=1}^{n} \phi(-L) f_i \nabla e_i \ , \ \phi : \{0, 1, 2, ...\} \longrightarrow \mathbf{R}.$$

(Note that $\phi(-L)\alpha$ is independent of the representation of α.) As special cases we define $L\alpha, J_k, T_t\alpha$ and $R_t\alpha$.

Definition 6.3. Suppose that $-\infty < t < \infty$ and $1 < p < \infty$

$$\| \alpha \|_{t,p} := \| R_{-t} \alpha \|_p .$$

Theorem 6.4. *$(\mathcal{P}^{1\otimes}, \| \ \|_{t,p})$ is a normed spaces, and $\| \alpha \|_{t,p} \uparrow$ as $t, p \uparrow$.*

Theorem 6.5. *$\| f \|_{t,r}$ and $\| \alpha \|_{t,p}$ are rotation invariant:*

$$\| Of \|_{t,p} = \| f \|_{t,p} \ , \ \| O\alpha \|_{t,p} = \| \alpha \|_{t,p} .$$

7. Generalized functions and generalized 1-forms

Definition 7.1. A linear functional $F : \mathcal{P} \rightarrow \mathbf{R}$ is called a *generalized function*, written $F \in \tilde{\mathcal{P}}$, and a linear functional $A : \mathcal{P}^{1\otimes} \rightarrow \mathbf{R}$ is called a *generalized 1-form*. The set of all generalized 1-forms, written $\tilde{\mathcal{P}}^{1\otimes}$, is a linear space with respect to the linear combinations defined in the obvious way.

For $f \in \mathcal{P}$ we define $F_f \in \tilde{\mathcal{P}}$ by

(1) $$F_f(g) := < f, g > \ , \ \forall g \in \mathcal{P}.$$

Then $f \mapsto F_f$ is a linear injection : $\mathcal{P} \to \tilde{\mathcal{P}}$. Hence we identify f with F_f, so that

(2) $$\mathcal{P} \subset \tilde{\mathcal{P}}$$

In view of

$$< f'f, g > = < f, f'g > \ , \ f', f, g \in \mathcal{P}$$
$$< \phi(-L)f, g > = < f, \phi(-L)g > \ , \ f, g \in \mathcal{P} \ , \ \phi : \{0, 1, 2, ...\} \to \mathbf{R},$$

we define $f'F, \phi(-L)F \in \tilde{\mathcal{P}}$ for $F \in \tilde{\mathcal{P}}$ as follows:

Definition 7.2. Let $f' \in \mathcal{P}$ and $\phi : \{0, 1, 2, ...\} \to \mathbf{R}$.

(3) $$(f'F)(g) := F(f'g) \ , \ \forall g \in \mathcal{P},$$

(4) $$(\phi(-L)F)(g) := F(\phi(-L)g) \ , \ \forall g \in \mathcal{P}.$$

As special cases of (4) we define $L, J_k, T_t, R_t : \tilde{\mathcal{P}} \to \tilde{\mathcal{P}}$.

For $\alpha \in \mathcal{P}^{1\otimes}$ we define $A_\alpha \in \tilde{\mathcal{P}}^{1\otimes}$ by
$A_\alpha(\beta) := < \alpha, \beta > \ , \ \forall \beta \in \mathcal{P}^{1\otimes}$. (See Definition 2.4 for $< \alpha, \beta >$.)

By identifying α with A_α , we obtain

(5) $$\mathcal{P}^{1\otimes} \subset \tilde{\mathcal{P}}^{1\otimes} .$$

In view of

$$< f\alpha, \beta > = < f, \alpha \cdot \beta >$$
$$< \gamma \cdot \alpha, f > = < \alpha, f\gamma >$$
$$< \nabla f, \beta > = < f, \nabla^* \beta >$$
$$< \nabla^* \alpha, f > = < \alpha, \nabla f >$$

(where $\alpha, \beta, \gamma \in \mathcal{P}^{1\otimes}$, $f \in \mathcal{P}$), we define $F \cdot \alpha, \gamma \cdot A, \nabla F \in \tilde{\mathcal{P}}^{1\otimes}$ and $\nabla^* A \in \tilde{\mathcal{P}}(F \in \tilde{\mathcal{P}}, A \in \tilde{\mathcal{P}}^{1\otimes}, \gamma, \beta \in \mathcal{P}^{1\otimes})$ as follows:

Definition 7.3.

$$(Fα)(β) = F(α \cdot β) \ , \ \forall β \in \mathcal{P}^{1\otimes} \tag{6}$$

$$(γ \cdot A)(f) = A(fγ) \ , \ \forall f \in \mathcal{P} \tag{7}$$

$$(\nabla F)(β) = F(\nabla^* β) \ , \ \forall β \in \mathcal{P}^{1\otimes} \tag{8}$$

$$(\nabla^* A)(f) = A(\nabla f) \ , \ \forall f \in \mathcal{P}. \tag{9}$$

Recalling the definition $LF(F \in \tilde{\mathcal{P}})$, we obtain

$$\begin{aligned}(LF)(f) &= F(Lf) = F(\nabla^* \nabla f) = (\nabla F)(\nabla f) \\ &= (\nabla^* \nabla F)(f) \ \text{ by (8) and (9),}\end{aligned}$$

so $L = \nabla^* \nabla$ on $\tilde{\mathcal{P}}$.

In view of the fact that $\| f \|_p = \sup\{| < f, g > | : g \in \mathcal{P}, \| g \|_q \leq 1\}, f \in \mathcal{P}$, where q is the cconjugate of p , we define $\| F \|_p$ for $F \in \tilde{\mathcal{P}}$ as follows:

Definition 7.4. $\| F \|_p := \sup\{|F(g)| : g \in \mathcal{P}, \| g \|_q \leq 1\}, F \in \mathcal{P}$.

Note that $0 \leq \| F \|_p \leq \infty$, $\| F + G \|_p \leq \| F \|_p + \| G \|_p$ and $\| cF \|_p = |c| \| F \|_p$ $(c \in \mathbf{R})$.

Extending the definition $\| f \|_{t,p} := \| R_{-t} f \|_p$ $(f \in \mathcal{P})$ we define

$$\| F \|_{t,p} := \| R_{-t} F \|_p \ , \ F \in \tilde{\mathcal{P}}.$$

$\| F \|_{t,p}$ satisfies the above-mentioned properties of $\| F \|_p$.

Similarly we define $\| A \|_p$ and $\| A \|_{t,p}$ $(1 < p < \infty, -\infty < t < \infty)$ as follows:

$$\| A \|_p := \sup\{|A(α)| \ , \ α \in \mathcal{P}^{1\otimes} \ , \ \| α \|_q \leq 1\} \ , \ A \in \tilde{\mathcal{P}}^{1\otimes},$$

and

$$\| A \|_{t,p} := \| R_{-t} A \|_p \ , \ A \in \tilde{\mathcal{P}}^{1\otimes}.$$

Definition 7.5. Suppose that $-\infty < t < \infty$ and $1 < p < \infty$.

$$\begin{aligned}\mathcal{D}_{t,p} &= \mathcal{D}_{t,p}(S) := \{F \in \tilde{\mathcal{P}} : \| F \|_{t,p} < \infty\} \\ \mathcal{D}^{1\otimes}_{t,p} &= \mathcal{D}^{1\otimes}_{t,p}(S) := \{A \in \tilde{\mathcal{P}}^{1\otimes} : \| F \|_{t,p} < \infty\}.\end{aligned}$$

Since it is easy to see that $\| F \|_{t,p} \uparrow$ as $t, p \uparrow \infty$ for $F \in \tilde{\mathcal{P}}$, we obtain

$$\mathcal{D}_{t,p} \downarrow \ \text{as } t, p \uparrow .$$

Similarly for $\mathcal{D}^{1\otimes}_{t,p}$.

Definition 7.6.

$$\mathcal{D} = \mathcal{D}(S) := \bigcap_{t,p} \mathcal{D}_{t,p} = \lim_{\substack{t\to\infty \\ p\to\infty}} \mathcal{D}_{t,p} \quad (\subset L_{\infty-} \subset L_p).$$

$$\tilde{\mathcal{D}} = \tilde{\mathcal{D}}(S) := \bigcup_{t,p} \mathcal{D}_{t,p} = \lim_{\substack{t\to-\infty \\ p\to 1}} \mathcal{D}_{t,p} \quad (\supset L_p \supset L_{\infty-}).$$

$F \in \mathcal{D}$ is called a *Malliavin function* on S, and $F \in \tilde{\mathcal{D}}$ is called a *Watanabe generalized function* on S. It is obvious that $\mathcal{P} \subset \mathcal{D}$.

Similarly we define *Malliavin 1-forms* and *Watanabe generalized 1-forms* to be elements of

$$\mathcal{D}^{1\otimes} = \bigcap_{t,p} \mathcal{D}^{1\otimes}_{t,p} \text{ and } \tilde{\mathcal{D}}^{1\otimes} := \bigcup_{t,p} \tilde{\mathcal{D}}^{1\otimes}_{t,p}$$

respectively. It is obvious that $\mathcal{P}^{1\otimes} \subset \mathcal{D}^{1\otimes} \subset \mathcal{D}^{1\otimes}_{t,p} \subset \tilde{\mathcal{D}}^{1\otimes}$.

It is easy to see that all concepts introduced in this section are rotation invariant.

8. Polynomial r-forms and generalized r-forms

In Sections 2 and 7 we defined polynomial 1-forms and generalized 1-forms. Extending these concepts, we define polynomial r-forms and generalized r-forms.

Let $\mathcal{T}$ be the polynomial vector fields defined in Section 2 and $\mathcal{T}^r$ the product of r copies of $\mathcal{T}$.

Definition 8.1. A map $\alpha : \mathcal{T}^r \to \mathcal{P}$ is called a *polynomial r-form*, written

$$\alpha \in \mathcal{P}^{r\otimes},$$

if α satisfies the following properties:

(i) $\alpha(V_1, V_2, ..., V_r)$ is *P-multilinear*, i.e. for every i, $\alpha(V_1, V_2, ..., V_r)$ is $\mathcal{P}$-linear in V_i, when $V_j, j \neq i$, are fixed.

(ii) *(Property of finite rank)* There exists a finite dimensional subspace $S_1 = S_1(\alpha) \subset S$ such that

$$\alpha(D_{h_1}, D_{h_2}, ..., D_{h_r}) = 0 \text{ if } h_i \in S \cap S^{\perp_1} \text{ for some i.}$$

We define $\mathcal{P}^{0\otimes} := \mathcal{P}$ by convention.

Definition 8.2. For $\alpha \in \mathcal{P}^{r\otimes}$ and $\beta \in \mathcal{P}^{s\otimes}$ we define their *tensor product* $\alpha \otimes \beta \in \mathcal{P}^{(r+s)\otimes}$ by

$$(\alpha \otimes \beta)(V_1, ..., V_r, V_{r+1}, ..., V_{r+s}) = \alpha(V_1, ..., V_r)\beta(V_{r+1}, ..., V_{r+s}), \ \forall\, V_i \in \mathcal{T}.$$

It is easy to check that the tensor product is associative, but not commutative. $\mathcal{P}^{r\otimes}$ is a $\mathcal{P}$-linear space.

Since $\nabla e \in \mathcal{P}^{1\otimes}$ for $e \in S$, we obtain

$$\nabla e_1 \otimes \nabla e_2 \otimes \ldots \otimes \nabla e_r \in \mathcal{P}^{r\otimes}, \quad e_i \in S.$$

Theorem 8.1. *Every $\alpha \in \mathcal{P}^{r\otimes}$ can be represented in the form*

$$\alpha = \sum_{i_1,\ldots,i_r=1}^{n} f_{i_1,\ldots,i_r} \nabla e_{i_1} \otimes \ldots \otimes \nabla e_{i_r}, \tag{1}$$

where $\{e_1, \ldots, e_n\}$ is ON in S and $f_{\ldots} \in \mathcal{P}$.

Definition 8.3. For $\alpha = \sum f_{i_1 i_2} \nabla e_{i_1} \otimes \nabla e_{i_2} \in \mathcal{P}^{2\otimes}$ and $\beta = \sum g_{j_1 j_2} \nabla h_{j_1} \otimes \nabla h_{j_2} \in \mathcal{P}^{2\otimes}$ we define the *dot product* $\alpha \cdot \beta \in \mathcal{P}$ by

$$\alpha \cdot \beta = \sum f_{i_1 i_2} g_{j_1 j_2} < e_{i_1}, h_{j_1} >< e_{i_2}, h_{j_2} > .$$

Similarly we define the *dot product* of two polynomial r-forms.

It is obvious that $\alpha \cdot \beta$ is $\mathcal{P}$-bilinear, $\alpha \cdot \beta = \beta \cdot \alpha$ and $\alpha \cdot \alpha \geq 0$ where the equality holds only for $\alpha = 0$. We define

$$|\alpha| := (\alpha \cdot \alpha)^{1/2} \in L_{\infty -}.$$

It is obvious that

$$(\nabla e_1 \otimes \ldots \otimes \nabla e_r) \cdot (\nabla h_1 \otimes \ldots \otimes \nabla h_r)$$
$$= < e_1, h_1 >< e_2, h_2 > \ldots < e_r, h_r > \quad (e_i, h_j \in S).$$

Define the inner product $< \alpha, \beta >$ for $\alpha, \beta \in \mathcal{P}^{r\otimes}$ and the p-norm $\| \alpha \|_p$ for $\alpha \in \mathcal{P}^{r\otimes}$ as follows:

$$< \alpha, \beta >= E(\alpha \cdot \beta) \text{ and } \| \alpha \|_p = \| \, |\alpha| \, \|_p .$$

If α is represented in the form (1), we define

$$\nabla \alpha = \sum \nabla f_{i_1 \ldots i_r} \otimes \nabla e_{i_1} \otimes \ldots \otimes \nabla e_{i_r}.$$

If $\alpha \in \mathcal{P}^{(r+1)\otimes}$, then α is represented in the form:

$$\alpha = \sum \alpha_{i_1 \ldots i_r} \otimes \nabla e_{i_1} \otimes \ldots \otimes \nabla e_{i_r}$$

where

$$\alpha_{i_1 \ldots i_r} \in \mathcal{P}^{1\otimes}, \quad e_i \in S.$$

We define

$$\nabla^* \alpha = \sum \nabla^* \alpha_{i_1 \dots i_r} \nabla e_{i_1} \otimes \dots \otimes \nabla e_{i_r}.$$

Then it is easy to check the following:

Theorem 8.2.

$$< \nabla \alpha, \beta > = < \alpha, \nabla^* \beta > \ , \ \alpha \in \mathcal{P}^{r\otimes} \ , \ \beta \in \mathcal{P}^{(r+1)\otimes}.$$

Define

$$L := \nabla^* \nabla.$$

The operators $\nabla : \mathcal{P}^{r\otimes} \to \mathcal{P}^{(r+1)\otimes}$, $\nabla^* : \mathcal{P}^{(r+1)\otimes} \to \mathcal{P}^{r\otimes}$ and $L : \mathcal{P}^{r\otimes} \to \mathcal{P}^{r\otimes}$ are called the *gradient operator*, the *divergent operator* and the *Ornstein-Uhlenbeck operator* respectively.

For the α of the form (1) we obtain

$$L\alpha = \sum L f_{i_1 \dots i_r} \nabla e_{i_1} \otimes \dots \otimes \nabla e_{i_r}.$$

Define $\phi(-L)$ for $\phi : \{0, 1, 2, \dots\} \to \mathbf{R}$ as follows:

$$\phi(-L)\alpha := \sum \phi(-L) f_{i_1 \dots i_r} \nabla e_{i_1} \otimes \dots \otimes \nabla e_{i_r}.$$

As special cases we define $T_t = e^{tL}$ and $R_t = (I - L)^{-t/2}$ on $\mathcal{P}^{r\otimes}$.

Theorem 8.3. *If* $t \geq 0$, *then*

$$\| T_t \alpha \|_p \leq \| \alpha \|_p \ , \ \| R_t \alpha \|_p \leq \| \alpha \|_p$$

for $\alpha \in \mathcal{P}^{r\otimes}$.

Theorem 8.4. (P. Meyer) *For* $r = 0, 1, 2, \dots, -\infty < t < \infty$ *and* $1 < p < \infty$ *there exists a positive number* $b_{r,t,p}$ *such that*

$$\| \nabla \alpha \|_{t,p} \leq b_{r,t,p} \| \alpha \|_{t+1,p} \text{ for } \alpha \in \mathcal{P}^{r\otimes}$$

which, combined with Theorem 8.2, implies

$$\| \nabla^* \alpha \|_{t,p} \leq b_{r,-t-1,q} \| \alpha \|_{t+1,p} \quad \alpha \in \mathcal{P}^{(r+1)\otimes}.$$

Let $\tilde{\mathcal{P}}^{r\otimes}$ be the linear functionals of $\mathcal{P}^{r\otimes}$. An element of $\mathcal{P}^{r\otimes}$ is called a *generalized r-form*. We define

$$L \ , \ \phi(-L) \ , \ T_t \ , \ R_t \ : \ \tilde{\mathcal{P}}^{r\otimes} \to \tilde{\mathcal{P}}^{r\otimes}$$

and

$$\nabla : \tilde{\mathcal{P}}^{r\otimes} \to \tilde{\mathcal{P}}^{(r+1)\otimes}, \quad \nabla^* : \tilde{\mathcal{P}}^{(r+1)\otimes} \to \tilde{\mathcal{P}}^{(r)}$$

in the same way as for the case $r-1$ (Definition 7). Also $\| A \|_{t,p}$ $(A \in \tilde{\mathcal{P}}^{r\otimes})$, $\mathcal{D}^{r\otimes}_{t,p}$, $\mathcal{D}^{r\otimes}$ and $\tilde{\mathcal{D}}^{r\otimes}$ are defined similarly.

From Theorems 8.3 and 8.4 we obtain the following:

Theorem 8.5. *If* $t \geq 0$, *then*

$$\| T_t A \|_p \leq \| A \|_p , \quad \| R_t A \|_p \leq \| A \|_p \quad \text{for } A \in \tilde{\mathcal{P}}^r .$$

Theorem 8.6. *For* $r = 0, 1, 2, ..., -\infty < t < \infty$ *and* $1 < p < \infty$ *we obtain*

$$\| \nabla A \|_{t,p} \leq b_{r,t,p} \| A \|_{t+1,p} \quad \text{for } A \in \tilde{\mathcal{P}}^{r\otimes}$$

and

$$\| \nabla^* A \|_{t,p} \leq b_{r,-t-1,q} \| A \|_{t+1,p} \quad \text{for } A \in \tilde{\mathcal{P}}^{(r+1)\otimes} .$$

These theorems imply that

$$\begin{aligned} T_t, R_t (t \geq 0) : \mathcal{D}^{r\otimes}_{t,p} &\to \mathcal{D}^{r\otimes}_{t,p} \\ \nabla : \mathcal{D}^{r\otimes}_{t,p} &\to \mathcal{D}^{(r+1)\otimes}_{t-1,p} \\ \nabla^* : \mathcal{D}^{(r+1)\otimes}_{t,p} &\to \mathcal{D}^{r\otimes}_{t-1,p} \\ L = \nabla^* \nabla : \mathcal{D}^{r\otimes}_{t,p} &\to \mathcal{D}^{r\otimes}_{t-2,p} \end{aligned}$$

and that these maps are linear and continuous. Hence $T_t, R_t, L : \mathcal{D}^{r\otimes} \to \mathcal{D}^{r\otimes}$ and

$$\nabla : \mathcal{D}^{r\otimes} \to \mathcal{D}^{(r+1)\otimes} , \quad \nabla^* : \mathcal{D}^{(r+1)\otimes} \to \mathcal{D}^{r\otimes},$$

and these maps are linear and continuous with respect to the quasinorm

$$\| A \| := \sum_{n=2}^{\infty} 2^{-n} (\| A \|_{n,n} \wedge 1).$$

9. The space $L^{r\otimes}_p$ that represents $\mathcal{D}^{r\otimes}_{0,p}$

By the duality between $(L_p, \| \ \|_p)$ and $(L_q, \| \ \|_q)$ we obtain the following:

Theorem 9.1. *The map* $f \mapsto F_f$ $(F_f(g) := < f, g >, g \in \mathcal{P})$ *gives an isomorphism from* $(L_p, \| \ \|_p)$ *to* $(\mathcal{D}_{0p}, \| \ \|_{0p})$.

By identifying f with F_f we obtain

$$L_p = \mathcal{D}_{0,p} \subset \underset{\sim}{\mathcal{D}}.$$

To extend this fact to $\mathcal{D}^{r\circledast}_{0,p}$ we first prove the following:

Theorem 9.2. *Let* $\underline{e} = \{e_1, e_2, ...\}$ *be an ONB in* S . *Then*

$A \in \mathcal{D}^{r\circledast}_{0,p}$ if and only if A is represented in the form:

$$A = \sum_{n_1 ... n_r} F_{n_1 ... n_r} \nabla e_{n_1} \otimes ... \otimes \nabla e_{n_r}$$

where the infinite sum is understood in the sense of $\| \ \|_p$*-convergence and* $\underline{F} = \{F_{n_1 ... n_r}\}$ *satisfies*

$$(1) \qquad \| (\sum_{n_1 ... n_r} F^2_{n_1 ... n_r})^{1/2} \|_p < \infty.$$

The coefficients $\{F_{n_1 ... n_r}\}$ *are determined by*

$$F_{n_1 ... n_r} = (\nabla e_{n_1} \otimes ... \otimes \nabla e_{n_2}) \cdot A$$

(see (7) of Definition 7.3), and it holds that $\| A \|_p$ *equals the left hand side of (1).*

Proof. We observe the case: $r = 1$ for simplicity of notation.

Suppose that $A \in \mathcal{D}^{1\circledast}_{0,p}$. Let

$$F_n := \nabla e_n \cdot A \ , \ i.e. \ F_n(f) = A(f \nabla e_n) \ , \ \forall f \in \mathcal{P}.$$

Then

$$|F_n(f)| \leq \| A \|_p \| f \nabla e_n \|_q = \| A \|_p \| f \|_q,$$

so

$$\| F_n \|_p \leq \| A \|_p < \infty.$$

Hence $F_n \in \mathcal{D}_{0,p} = L_p$ and

$$F_n(f) = E(F_n f).$$

Observe that

$$|A(\sum_{n=1}^{N} f_n \nabla e_n)| \leq \| A \|_p \| \sum_{n=1}^{N} f_n \nabla e_i \|_q \ , \ f_n \in \mathcal{P}.$$

$$|E(\sum_{n=1}^{N} F_n f_n)| \leq \| A \|_p \| (\sum_{n=1}^{N} f_n{}^2)^{1/2} \|_q \ .$$

Since $\mathcal{P}$ is dense in L_q , this implies that

$$\| (\sum_{n=1}^{N} F_n{}^2)^{1/2} \|_p \leq \| A \|_p \ .$$

Letting $N \to \infty$, we obtain

$$\| (\sum_{n=1}^{\infty} F_n{}^2)^{1/2} \|_p \leq \| A \|_p \quad (< \infty). \tag{2}$$

Observe that

$$\begin{aligned} |A(\sum_{n=1}^{N} f_n \nabla e_n)| = |E(\sum_{n=1}^{N} F_n f_n)| &\leq \| (\sum_{n=1}^{N} F_n^2)^{1/2} \|_p \| (\sum_{n=1}^{N} f_n^2)^{1/2} \|_q \\ &\leq \| (\sum_{n=1}^{\infty} F_n^2)^{1/2} \|_p \| \sum_{n=1}^{N} f_n \nabla e_n \|_q \ . \end{aligned}$$

Since $\{ \sum_{n=1}^{N} f_n \nabla e_n \}_{N, f_n}$ is dense in $(\mathcal{P}^{1\otimes}, \| \ \|_q)$, this implies that

$$\| A \|_p \leq \| (\sum_{n=1}^{\infty} F_n{}^2)^{1/2} \|_p \tag{3}$$

which, combined with (2), implies that

$$\| A \|_p = \| (\sum_{n=1}^{\infty} F_n{}^2)^{1/2} \|_p \ . \tag{4}$$

Applying this equality to $R_N = A - \sum_{n=1}^{N} F_n \nabla e_n \in \mathcal{D}_{0,p}^{r\otimes}$ and noting that $R_N \cdot \nabla e_m = 0 (m \leq N)$ and $= F_m \, (m > N)$, we obtain

$$\| R_N \|_p = \| (\sum_{n=N+1}^{\infty} F_n{}^2)^{1/2} \|_p \longrightarrow 0 \quad (n \to \infty),$$

i.e. $A = \sum_{n=1}^{\infty} F_n \nabla e_n$.

Suppose that $\| (\sum_{n=1}^{\infty} F_n{}^2)^{1/2} \|_p < \infty$. Let

$$A_N := \sum_{n=1}^{N} F_n \nabla e_n .$$

Then $A_M - A_N \in \mathcal{D}_{0p}^{1\otimes}$ and $(A_M - A_N) \cdot \nabla e_n = F_n$ for $N < n \leq M$. Applying (4) to $A_M - A_N$, we obtain

$$\| A_m - A_N \|_p = \| (\sum_{n=N+1}^{M} F_n{}^2)^{1/2} \|_p \longrightarrow 0 \quad (M > N \to \infty).$$

Since $(\mathcal{D}_{0p}^{1\otimes}, \| \ \|_p)$ is a Banach space, we can find $A \in \mathcal{D}_{0,p}^{1\otimes}$ such that $\| A - A_N \|_p \to 0$ as $N \to \infty$, i.e.

$$\sum_{n=1}^{\infty} F_n \cdot \nabla e_n = A \in \mathcal{D}_{0,p}^{1\otimes}. \quad \square$$

We denote $\{A \in \mathcal{D}_{0,p}^{r\otimes} : A \cdot \nabla h_1 \otimes ... \otimes \nabla h_r \in \mathbf{R}, \forall h_i \in S\}$ by $H_p^{(r)}$ for the moment. By the last theorem $A \in H_p^{(r)}$ if and only if A is represented in the form

$$A = \sum c_{n_1 \dots n_r} \nabla e_{n_1} \otimes ... \otimes \nabla e_{n_r}, \tag{5}$$

where $\{e_n\}$ is an ONS in S , $c_{n_1 \dots n_r} \in \mathbf{R}$ and

$$\sum c^2_{n_1 \dots n_r} < \infty.$$

Hence $H_p^{(r)}$ is independent of p , so we denote it simply by $H^{(r)}$. Hence $H^{(r)}$ is a Hilbert space with the inner product and the norm:

$$< A, B >_{H^{(r)}} = \sum c_{n_1 \dots n_r} d_{n_1 \dots n_r} , \quad \| A \|_{H^{(r)}} = (< A, A >_{H^{(r)}})^{1/2}$$

when A and B are of the form (5) with coefficients $\{c...\}$ and $\{d...\}$ respectively. Then $\{\nabla e_{n_1} \otimes ... \otimes \nabla e_{n_r}\}$ is an ONB in $H^{(r)}$. Denote $H^{(1)}$ simply by H . Then (5) shows that $H^{(r)}$ is the tensor product of r copies of H , so we denote $H^{(r)}$ by $H^{r\otimes}$.

Definition 9.1.

$$\begin{aligned} L_0^{r\otimes} &= L_0^{r\otimes}(S) \\ &:= \textit{the } H^{r\otimes} - \text{valued function on } \Omega \text{ measurable } \sigma[S]/\mathcal{B}(H^{r\otimes}), \\ \| \Phi \|_p &:= (\int_\Omega \| \Phi(\omega) \|_{H^{r\otimes}}^p P(d\omega))^{1/2} , \quad \Phi \in L_0^{r\otimes}, \\ L_p^{r\otimes} &= L_0^{r\otimes}(S) \\ &:= \{\Phi \in L_0^{r\otimes} : \| \Phi \|_p < \infty\} , \quad 1 \leq p < \infty. \end{aligned}$$

For $\Phi \in L_p^{r\otimes}$ we set

$$A_\Phi(\beta) = \int_\Omega < \Phi(\omega), \beta(\omega) >_{H^{r\otimes}} P(d\omega)$$

where $\beta(\omega) := \sum f_{i_1 \dots i_r}(\omega)\nabla e_{i_1} \otimes \dots \otimes \nabla e_{i_r} \in H^{r\otimes}$ for $\beta = \sum f_{i_1 \dots i_r} \nabla e_{i_1} \otimes \dots \otimes \nabla e_{i_r} \in \mathcal{P}^{r\otimes}$. Then $A_\Phi \in \mathcal{P}^{\sim r\otimes}$. We can prove that $\Phi \to A_\Phi$ is linear and

$$\| A_\Phi \|_p = \| \Phi \|_p .$$

Hence $\Phi \to A_\Phi$ gives an isomorphism from $(L_p^{r\otimes}, \| \ \|_p)$ to $(\mathcal{D}_{0p}^{r\otimes}, \| \ \|_{0p} = \| \ \|_p)$.

From now on we identify Φ with A_Φ , so

$$L_p^{r\otimes} = \mathcal{D}_{0p}^{r\otimes} .$$

Definition 9.2. Suppose that $p^{-1} = p_1^{-1} + p_2^{-1}$ where $0 < p_1, p_2, p < \infty$.

(i) For $A \in L_{p_1}^{r\otimes}$ and $B \in L_p^{r\otimes}$ we define $A \cdot B \in L_p$ by

$$(A \cdot B)(\omega) = < A(\omega), B(\omega) > H^{r\otimes} .$$

(ii) For $A \in L_{p_1}^{r_1\otimes}$ and $B \in L_{p_2}^{r_2\otimes}$ we define $A \otimes B \in L_p^{r\otimes}$ $(r = r_1 + r_2)$ by

$$\begin{aligned} &< A(\omega) \otimes B(\omega), \alpha \otimes \beta > H^{r\otimes} \\ &=< A(\omega), \alpha >_H r_1 \otimes < B(\omega), \beta > H^{r_2\otimes} \end{aligned}$$

for $\alpha = \nabla h_1 \otimes \dots \otimes \nabla h_{r_1}$ and $\beta = \nabla h_{r_1+1} \otimes \dots \otimes \nabla h_{r_1+r_2}$ $(h_1, h_2, \dots, h_{r_1+r_2} \in S)$.

Note that

(i) $|(A \cdot B)(\omega)| \le \| A(\omega) \|_{H^{r_1\otimes}} \| B(\omega) \|_{H^{r_2\otimes}}$, so

$$\| A \cdot B \|_p \le \| A \|_{p_1} \| B \|_{p_2} ;$$

(ii) $\| (A \otimes B)(\omega) \|_{H^{r\otimes}} = \| A(\omega) \|_{H^{r_1\otimes}} \| B(\omega) \|_{H^{r_2\otimes}}$, so

$$\| A \otimes B \|_p \le \| A \|_{p_1} \| B \|_{p_2} .$$

10. Integral representation of $L_p^{r\otimes}$

Let $\Lambda = (\Lambda, \mathcal{B}(\Lambda), m)$ be a σ-finite measure space and let

$$\mathcal{B}_f(\Lambda) := \{E \in \mathcal{B}(\Lambda) : m(E) < \infty\}.$$

Suppose that $\{e_E \in S, E \in \mathcal{B}_f(\Lambda)\}$ satisfies

(1) $$< e_E, e_{E'} >= m(E \cap E').$$

Then if $E_n \in \mathcal{B}_f(\Lambda)$, $n = 1,2,...$ are disjoint and if $E := \bigcup_n E_n \in \mathcal{B}_f(\Lambda)$, then

$$e_E = \sum_{n=1}^{\infty} e_{E_n} \text{ in } (S, \| \ \|_2).$$

In view of this fact the system $\{e_E, E \in \mathcal{B}_f(\Lambda)\}$ satisfying (1) is called an *S-valued measure* on Λ.

Let $\{e_E\}$ be an S-valued measure on Λ. Then $\nabla e_E \in H$ and

$$< \nabla e_E, \nabla e_{E'} >_H = < e_E, e_{E'} > = m(E \cap E').$$

Hence $\{\nabla e_E, E \in \mathcal{B}_f(\Lambda)\}$ is regarded an H-valued measure on Λ.

Let $\Lambda^r := (\Lambda^r, \mathcal{B}(\Lambda)^r, m^r)$. Since

$$\begin{aligned}
&< \nabla e_{E_1} \otimes ... \otimes \nabla e_{E_r}, \nabla e_{E'_1} \otimes ... \otimes \nabla e_{E'_r} > \\
&= m(E_1 \cap E'_1)...m(E_r \cap E'_r) \\
&= m((E_1 \times .... \times E_r) \cap (E'_1 \times ... \times E'_r))
\end{aligned}$$

the $H^{r\otimes}$-valued set function $\xi(E_1 \times ... \times E_r) := \nabla e_{E_1 \otimes ... \otimes \nabla e_{E_r}}$ can be extended to an $H^{r\otimes}$-valued measure on Λ^n. Hence we can define the integral:

$$\int_{\Lambda^n} \varphi(\lambda_1, \lambda_2, ..., \lambda_r) \nabla e_{d\lambda_1} \otimes ... \otimes \nabla e_{d\lambda_r} \in H^{r\otimes}$$

for $\varphi \in L_2(\Lambda^n)$ in the obvious way.

Suppose that $F_{\lambda_1...\lambda_r} \in L_p$ for $\underline{\lambda} = (\lambda_1, \lambda_2, ..., \lambda_r) \in \Lambda^r$ and that $F_{\underline{\lambda}}(\omega)$ is $\mathcal{B}(\Lambda)^r \otimes \sigma[S]$-measurable in $(\underline{\lambda}, \omega) \in \Lambda^r \times \Omega$. If

$$\int_\Omega \left(\int_\Lambda |F_{\underline{\lambda}}(\omega)|^2 m^r(d\underline{\lambda}) \right)^{p/2} P(d\omega) < \infty,$$

then

$$\int_\Lambda |F_{\underline{\lambda}}(\omega)|^2 m^r(d\underline{\lambda}) < \infty \text{ a.s.},$$

i.e.

$$F_{\underline{\lambda}}(\omega) \in L_2(\Lambda^r) \text{ (as a function of } \underline{\lambda}) \text{ a.s.}$$

Hence

$$\Phi(\omega) := \int_{\Lambda^r} F_{\lambda_1...\lambda_r}(\omega) \nabla e_{d\lambda_1} \otimes ... \otimes \nabla e_{d\lambda_r} \in H^{r\otimes} \text{ a.s.}$$

and $\| \Phi \|_p = \| \| \Phi \|_{H^{r\otimes}} \|_p < \infty$. This implies that

$$\Phi \in L_p^{r\otimes}.$$

Theorem 10.1. *If $F_{\lambda_1...\lambda_r} \in \mathcal{D}, \forall \lambda_1...\lambda_r$, then $\Phi \in \mathcal{D}^{r\otimes}$ and*

(i) $\nabla\Phi = \int_{\Lambda^r} \nabla F_{\lambda_1\ldots\lambda_r} \otimes \nabla e_{d\lambda_1} \otimes \ldots \otimes \nabla e_{d\lambda_r} \in \mathcal{D}^{(r+1)\otimes}\ (r\geq 0)$
(ii) $\nabla^*\Phi = -\int_{\Lambda^r}(\nabla F_{\lambda_1\ldots\lambda_r} \cdot \nabla e_{d\lambda_1} - F_{\lambda_1\ldots\lambda_r} e_{d\lambda_1})\nabla e_{d\lambda_2} \otimes \ldots \otimes \nabla e_{d\lambda_r}$

$\in \mathcal{D}^{(r-1)\otimes}\ (r \geq 1)$.
Note that $\nabla F \cdot \nabla e = <\nabla F, \nabla e>_{H^{1\otimes}}$.

11. Stochastic integrals and divergent operators

Let $\{B_t(\omega), 0 \leq t < \infty\}$ be a Wiener process on $\Omega = (\Omega, \mathcal{F}, P)$. Then the closed linear span of $B_t, 0 \leq t < \infty$ in $L_2(\Omega)$ is a Segal system, which is denoted by S . Let S_t be the closed linear span of B_s , $0 \leq s \leq t$ in $L_2(\Omega)$. S_t is a Segal subsystem of S . It is obvious that $\mathcal{D}(S)$. In fact $\mathcal{D}(S_t)$ is the set of all elements of S measurable $\sigma[B_s, s \leq t]$. Then $\mathcal{D}(S_t)$ is a closed linear subset of

$$B_E := \int_0^\infty 1_{E}(t) dB_t \quad \text{(Wiener integral)}$$

$E \in \mathcal{B}[0,\infty), |E| < \infty$ ($|\ |$= Lebesque measure).

is an S-valued measure on $[0,\infty)$. Hence $\{\nabla B_E\}_E$ is an H-valued measure on $[0,\infty)$. Applying Theorem 10.1 to $\{\nabla B_E\}_E$ we obtain

$$\begin{aligned}\nabla^*\Big(\int_0^\infty F_t \nabla B_{dt}\Big) &= -\int_0^\infty (\nabla F_t \cdot \nabla B_{dt} - F_t B_{dt}) \\ &= -\int_0^\infty \nabla F_t \cdot \nabla B_{dt} + \int_0^\infty F_t B_{dt}.\end{aligned}$$

Theorem 11.1. *(Gaveau-Trauber) If* $F_t \in \mathcal{D}(S_t)$ *, then* $\nabla F_t \cdot \nabla B_{dt} = 0$ *, so*

$$\int_0^\infty F_t B_{dt} = \nabla^*\Big(\int_0^\infty F_t \nabla B_{dt}\Big),$$

where the left hand side is coincident with the stochastic integral $\int_0^\infty F_t dB_t$.

References

[1] P. Malliavin: Stochastic calculus of variation and hypo-elliptic operators, *Proc. Int. Symp. S.D.E.*, Kyoto (1976), Kinokuniya, 1978, 195-263.

[2] D.W. Stroock: The Malliavin calculus and its applications to second order parabolic differential operators, I, II, *Math. System Theory 14*, 1981, 25-65, 141-171.

[3] S. Watanabe: Stochastic differential equations and Malliavin calculus, Lecture Note Series 73, *Tata Institute of Fundamental Research*, Bombay, 1984.

[4] I. Shigekawa: *Malliavin calculus* (in Japanese).

[5] K. Itô: Malliavin's C^∞ functionals of a centered Gaussian system, *IMA Preprint Series 327*, Univ. of Minnesota, 1987.

[6] K. Itô: Malliavin calculus on a Segal space, *Inter Conference in honor of P. Lévy*, Paris, 1987.

Invited lectures

Thoughts on the modelling and identification of random processes and fields subject to possible long-range dependence

C. C. Heyde and Roger Gay

1. Introduction

The recognition of stationary random processes and random fields subject to long-range dependence is a problem of particular contemporary scientific importance. This is because such processes typically exhibit trends or periodicities which appear persistent but are eventually ephemeral. For example, to distinguish between a stationary short-range dependence process with trend and a stationary long-range dependence process as a model for CO_2 or hydrofluorocarbons in the atmosphere is of particular significance in connection with the recently commenced International Geosphere-Biosphere Programme (IGBP) of the International Council of Scientific Unions (ICSU).

The phenomenon of long-range dependence has now been observed in widely differing areas such as hydrology, geophysics and agriculture. Early contributors were Fairfield-Smith (1938) and Hurst (1951) in the contexts of crop yields (as a function of area) and river flows respectively. A good perspective on circumstances under which long-range dependence occurs is given by Hampel (1987).

The discrimination between models involving short- and long-range dependence is crucial since the latter behaviour ordinarily has drastic consequences for statistical inference, in particular for the prescription of confidence intervals. For example, instead of the asymptotic normality of $n^{\frac{1}{2}}(\bar{X}_n - \mu)$ for the mean $\bar{X}_n$ of a sample of size n from a stationary process with short-range dependence, we would rather typically have normality of $n^{1-H}(\bar{X}_n - \mu)$ for a certain H, $\frac{1}{2} < H < 1$, in the long-range dependence case.

In this paper we shall discuss a rich class of models, involving both short- and long-range dependence, for which there is a unified approach to the corresponding parameter inference. The work was motivated by a problem posed by the Aeronautical Research Laboratories (in Australia) necessitating the building of a broad class of topographical landscape models, ranging from "smooth" to "rough" in character, with straightforward identifiability properties.

2. Dependence

To fix ideas we consider the random process case. Let $\{X_t,\ t = 1, 2, ...\}$ be a

Probability Theory

second-order stationary process with covariances

$$\gamma_k = cov(X_t, X_{t+k})$$

and spectral density

$$f(\omega) = \frac{1}{2\pi} \sum_{k=-\infty}^{\infty} \gamma_k e^{-ik\omega}, \quad -\pi < \omega < \pi.$$

We also consider the block means process $\{X_t^{(m)}\}$ given by

$$X_t^{(m)} = (X_{tm-m+1} + \ldots + X_{tm})/m$$

and the Allen variance

$$V_m = \text{var } X_t^{(m)}.$$

Then, the dichotomy between short- and long-range dependence can be characterized by the (essentially equivalent) criteria given in the following table (e.g. Cox (1984)).

Short-range dependence	Long-range dependence
(i) $\sum_0^\infty \gamma_k$ converges	(i) $\sum_0^\infty \gamma_k$ diverges
(ii) $f(0)$ finite	(ii) $f(\omega)$ singular at $\omega = 0$
(iii) $mV_m \to$ finite const. as $m \to \infty$	(iii) $mV_m \to \infty$ as $m \to \infty$

For short-range dependence the averaged process tends to second order pure noise, while for long-range dependence the averaged process typically has a non-degenerate correlation structure as $m \to \infty$ and the variance of the mean decreases more slowly than the reciprocal of the sample size. Similar results hold for random fields.

3. Model building

Our approach to the modelling problem is to assume that X is a continuous time process, although possibly observed on a lattice, and to specify a class of transfer functions which can be applied to white noise which encapsulates a satisfactory range of low frequency behaviour including both short- and long-range dependence. For futher details see Gay and Heyde (1990a).

To allow for the spectral density $f(\omega)$ to diverge at $\omega = 0$ we can introduce the operational stochastic differential equation

$$(D + \alpha)^\nu X(t) = e(t), \quad \nu > 0, \tag{1}$$

where $e(t)$ denotes white noise, $D = d/dt$ and

$$(D+\alpha)^{\nu} = \sum_{j=0}^{\infty} \frac{\nu(\nu-1)\ldots(\nu-j+1)}{j!} \alpha^{\nu-j} D^{j}.$$

This has a stationary solution with spectral density

(2) $$f_{\alpha,\omega}(\omega) = \sigma^2/2\pi(\alpha^2+\omega^2)^{\nu}$$

which exhibits short-range dependence if $\alpha > 0$. However, if $\alpha = 0$ we have long-range dependence. When $0 < \nu < \frac{1}{2}$, (2) is not the spectral density of a stationary process, but of a self-similar stationary increment process of order $\nu + \frac{1}{2}$. It is, however, very closely approximated by a (continuous) stationary process with spectral density

$$f_Z(\omega) = 4\sigma^2(2\pi)^{-1}(sin^2\omega/2)\omega^{-2-2\nu}, \quad \omega \in (-\infty, \infty), \quad 0 < \nu < \frac{1}{2},$$

derived from

$$Z(t) = \int_{t-\frac{1}{2}}^{t+\frac{1}{2}} Y(u)du,$$

where the process $\{Y(t), -\infty < t < \infty\}$ has stationary increments and spectral density (2). A closely corresponding version for Z observed only on a periodic lattice has spectral density

$$f_d(\omega) = 2(1-cos\,\omega)\sigma^2(2\pi)^{-1} \sum_{u=-\infty}^{\infty} |\omega + 2\pi u|^{-2-2\nu}, \quad \omega \in [-\pi, \pi],$$

that of discrete fractional Gaussian noise.

The family of models encompassed by (2) can be augmented by the rational transfer functions of ARMA type processes to cover more complicated features in the higher frequency part of the spectrum. If we replace (1) by the equation

(3) $$Q(D)(D+\alpha)^{\nu} X(t) = P(D)e(t)$$

for polynomials, P, Q, then we obtain the family of spectral densities of the form

(4) $$\sigma^2|P(i\omega)|^2/2\pi|Q(i\omega)|^2(\alpha^2+\omega^2)^{\nu}.$$

This family answers the need expressed by Cox (1984) for "a family of Gaussian processes reducing to the self-similar processes as a special case and having sufficient additional parameters to accommodate fairly general short-range dependence".

For random fields of dimension $n > 1$ we use the stochastic operational Laplace equation

(5) $$(\nabla^2 - \beta^2)^{\nu} X(t) = e(t), \quad \nu > 0, \beta \in \mathbf{R},$$

where $e(t)$ is a white noise field, ∇^2 is the Laplacian, and

$$(\nabla^2 - \beta^2)^\nu = (-1)^\nu \sum_{j=0}^{\infty} \binom{\nu}{j} (-\nabla^2)^j \beta^{2(\nu-j)}.$$

Then, a stationary isotropic solution to equation (5) has spectral density

$$f_{\beta,\nu}(\omega) = (\sigma^2/(2\pi)^n)/(|\omega|^2 + \beta^2)^{2\nu} \tag{6}$$

with $|\omega|^2 = \omega_1^2 + \ldots + \omega_n^2$.

We have here used an isotropic version but some anisotropy can be injected by putting

$$\omega = (\omega_1/a_1, \ldots, \omega_n/a_n)$$

and rotation of the axes is also possible.

For $\beta^2 > 0$ the process X in (5) exhibits short-range dependence. However, for $(n-1)/4 < \nu < n/4$,

$$f_{0,\nu}(\omega) = \sigma^2 (2\pi)^{-n} |\omega|^{-4\nu}$$

is the n-dimensional spectral density of an isotropic self-similar random field $Y(t)$, say, with stationary increments. Also, the process

$$Z(t) = \int_{t_1 - \frac{1}{2}}^{t_1 + \frac{1}{2}} \cdots \int_{t_n - \frac{1}{2}}^{t_n + \frac{1}{2}} Y(u_1, \ldots, u_n) du_1 \ldots du_n,$$

which has spectral density

$$f_Z(\omega) = \sigma^2 \pi^{-n} \{\sum_{i=1}^{n} \omega_i^{-2} \sin^2(\omega_i/2)\} (\omega_1^2 + \ldots + \omega_n^2)^{-2\nu}$$

is stationary and asymptotically isotropic (when $a_1 = a_2 = \ldots = a_n$) with asymptotic covariance $C(\tau) \sim c_2 \tau^{4\nu - n}$ as $\tau \to \infty$, c_2 constant (e.g. (Sinai (1976)). Further, the same asymptotic covariance is also possessed by the closely corresponding process Z_t, say, on the (unit) lattice with spectral density

$$f_d(\omega) = \sum_{u_1=-\infty}^{\infty} \cdots \sum_{u_n=-\infty}^{\infty} f_Z(\omega_1 + 2\pi u_1, \ldots, \omega_n + 2\pi u_n).$$

As with the one dimensional case, the family of models encompassed by (6) can usefully be augmented by rational transfer functions to cover more complicated features in the higher frequency part of the spectrum. We could thus prescribe, in extension of the family discussed by Veccia (1988), spectral densities of the form

$$\sigma^2 \prod_{j=1}^{p} \left| |\omega|^2 + \theta_j \right|^{2n_j} / (2\pi)^n (|\omega|^2 + \beta^2)^{2\nu} \prod_{i=1}^{q} (|\omega|^2 + \phi_i)^{2m_i} \tag{7}$$

where n_j and m_i are positive integers satisfying $\sum m_i \geq \sum n_j + 1$, while θ_j, $1 \leq j \leq p$ are real and ϕ_i, $1 \leq i \leq q$, are positive. This family covers a comprehensive range of short- and long-range behaviour.

4. Inference

Inference for the models (4) and (7) can be usefully based on the periodogram and then does not involve specific distributional assumptions for the random process case; a full discussion of the results, including the random field case, can be found in Gay and Heyde (1990b).

Suppose that a sample $\{X_t,\ 0 \leq t \leq T\}$ is given from a stationary process with spectral density $f(\omega, \theta)$, the distribution of X depending on θ (e.g. $\theta = (\alpha, \nu)'$ for (2)). Let $I_T(\omega)$ denote the periodogram. Then, it is ordinarily the case that

$$(f(\omega,\theta))^{-1} I_T(\omega) \xrightarrow{d} \chi_2^2 \tag{8}$$

as $T \to \infty$, i.e. convergence in distribution to a chi-squared distribution with two degrees of freedom, for both short- and long-range dependence.

Smoothing of the periodogram is necessary for inference since the convergence in distribution result (8) is not amenable to direct use. Estimation is via the class of estimating functions

$$G_T(\theta) = \int_{-\pi}^{\pi} A(\omega,\theta)[I_T(\omega) - EI_T(\omega)]d\omega \tag{9}$$

for vector smoothing functions $A(\omega, \theta)$ (having the dimension of θ) such that

$$\int_{-\pi}^{\pi} A'(\omega,\theta)A(\omega,\theta)d\omega < \infty, \quad A(\omega,\theta) = A(-\omega,\theta).$$

To find an estimator $\hat{\theta}$ of θ we solve the estimating equation

$$G_T(\hat{\theta}) = 0.$$

The asymptotic quasi-likelihood estimator for θ, which provides minimum size asymptotic confidence zones for θ, is obtained from the choice for A of

$$A^*(\omega,\theta) = \dot{f}(\omega,\theta)/f^2(\omega,\theta),$$

the dot denoting vector differentiation with respect to θ. This procedure is asymptotically equivalent to choosing θ to minimize

$$\int_{-\pi}^{\pi} [\log f(\omega,\theta) + \frac{I_T(\omega)}{f(\omega,\theta)}]d\omega,$$

often called the Whittle procedure, which is motivated by approximate likelihood considerations under a Gaussian distributional assumption for X. For a discussion of the asymptotic quasi-likelihood methodology see Heyde and Gay (1989) and references therein. A particular virtue of the quasi-likelihood approach is that it provides a convenient framework for estimation without the Gaussian assumption; the same asymptotic theory applies.

Suppose that $\{X_t\}$ is a purely non-deterministic process with finite fourth moment and that $f(\omega,\theta) \in L_1$ over $[-\pi,\pi]$. Then,

$$T^{\frac{1}{2}} \int_{-\pi}^{\pi} \dot{f}(\omega,\theta) f^{-2}(\omega,\theta)[I_T(\omega) - EI_T(\omega)]d\omega \tag{10}$$

converges in distribution to the multivariate normal law $MVN(0,W)$ where

$$W = 4\pi \int_{-\pi}^{\pi} \dot{f}(\omega,\theta)(\dot{f}(\omega,\theta))' f^{-2}(\omega,\theta)d\omega$$

and, under minor additional regularity conditions on f alone, Taylor expansion of (10) leads to

$$T^{\frac{1}{2}}(\hat{\theta} - \theta) \xrightarrow{d} MVN(0,W^{-1}). \tag{11}$$

As noted above, the quasi-likelihood theory assures us that the asymptotic confidence zones given by (11) are best amongst the family of estimating functions given by (9). It should, furthermore, be emphasized that the results hold under circumstances of either short- or long-range dependence. Previous discussions of results like (11) had imposed conditions of short-range dependence (e.g. Rosenblatt (1985), Chapter IV) or an underlying Gaussian distribution (e.g. Fox and Taqqu (1986), and Dahlhaus (1987)). In the case of (2), for example, estimation of $\theta = (\alpha,\nu)'$ suggests the nature of dependence of the process X_t.

References

[1] D. R. Cox, Long-range dependence : a review. In *Statistics an Appraisal*, eds. H. A. David and H. T. David, Iowa State Univ. Press, Ames, (1984), 55-74.

[2] R. Dahlhaus, Efficient parameter estimation for self-similar processes, Preprint, (1987).

[3] H. Fairfield-Smith, An empirical law describing heterogeneity in yields of agricultural crops, *J. Agric. Sci.* **28**(1938), 1-23.

[4] R. Fox & M. S. Taqqu, Large sample properties of parameter estimates for strongly dependent stationary Gaussian time series, *Ann. Statist.*, **14**(1986), 517-532.

[5] R. Gay & C. C. Heyde, On a class of random field models which allows long-range dependence. *Biometrika*, **77**(1990), 401-403.

[6] R. Gay & C. C. Heyde, Asymptotics for the smoothed periodogram and parameter estimation for random processes and random fields without Gaussian assumptions and subject to possible long-range dependence, manuscript (1990b).

[7] F. Hampel, Data analysis and self-similar processes. *Bull. Int. Statist. Inst.*, **52**(1987), Book 4, 235-254.

[8] C. C. Heyde & R. Gay, On asymptotic quasi-likelihood estimation. *Stochastic Process. Appl.*, **31**(1989), 223-236.

[9] W. E. Hurst, Long-term storage capacity of reservoirs. *Trans. Amer. Soc. Civil Engineers*, **116**(1951), 770-808.

[10] M. R. Rosenblatt, *Stationary Sequences and Random Fields*, Birkhäuser, Basel (1985).

[11] Ya. G. Sinai, Self-similar probability distributions. *Theory Probab. Appl.*, **21**(1976), 64-80.

[12] A. V. Veccia, Estimation and model identification for continuous spatial processes. *J. Roy. Statist. Soc. Ser. B*, **50**(1988), 297-312.

White noise and Gaussian random fields

Takeyuki Hida

Introduction

We shall discuss Gaussian random fields indexed by a manifold running through an Euclidean space. Our approach to such fields is in line with white noise analysis and the present note uses, as usual, two basic tools in the white noise analysis: namely, one is the class of generalized white noise functionals as well as their integral representations (see, e.g. [4], [8], [10]) where the parameter of white noise is multi-dimensional, and the other is the infinite dimensional rotation group (see [5], [6], [7]). With these tools we can analyze Gaussian random fields by using the representation theory in terms of white noise, and then we shall proceed to discussions of them by applying the variational calculus in the case where the parameter is taken to be a manifold in $\mathbf{R}^d$ and varies or is distorted within a certain class of manifolds.

Changing the parameter manifold can be carried out by the action of the infinite dimensional rotation group, so that our approach may be thought of as a *harmonic analysis* arising from the rotation group, although what we discuss is far from the general theory.

Having obtained the variation, we are given infinitesimal random variables, some of which are expressed as white noise integrals with singular kernels functions. In order to give rigorous mathematical description to such infinitesimal random variables, we need some more background which will be provided in Section 2.

We shall take two classes of basic Gaussian random fields that are depending on plane curve in Section 3. There we shall show that white noise can be squeezed in from the given random field with the help of its variation and it can be regarded as the *innovation* of the field. This approach seems to be a generalization of the result that was obtained by P. Lévy [2] in the one-dimensional parameter case, i.e. the case of ordinary Gaussian processes.

We have hopes that more general classes of Gaussian random fields can also be discussed within our setting.

1. Background

An $\mathbf{R}^d$-parameter white noise is a generalized Gaussian process with the characteristic functional

(1) $$C(\xi) = \exp[-\frac{1}{2} \| \xi \|^2], \quad \xi \in E,$$

where E is a function space being a member of a Gel'fand triple:

(2) $$E \subset L^2(\mathbf{R}^d) \subset E^*, \quad E^* \text{ being the dual space of } E.$$

Probability Theory

The basic space E is usually taken to be a nuclear space such as the Schwartz space S, however the Sobolev space of suitable order is also fitting for the present purpose.

Associated with this $C(\xi)$ is a probability measure μ on E^* such that

$$C(\xi) = \int_{E^*} \exp[i < x, \xi >] d\mu(x). \tag{3}$$

There is given a complex Hilbert space $(L^2) \equiv L^2(E^*, \mu)$, which admits a direct sum decomposition, called Wiener-Itô decomposition:

$$(L^2) = \sum_{n=0}^{\infty} \oplus \mathcal{H}_n. \tag{4}$$

The S-transform of an (L^2)-functional is defined by

$$\varphi \longrightarrow U_\varphi(\xi) = U(\xi) = \int_{E^*} \varphi(x+\xi) d\mu(x), \quad \varphi \in (L^2),$$

and it establishes an isomorphism

$$(L^2) \simeq \sum \oplus \hat{L}^2(\mathbf{R}^{nd}, n!du^n), \tag{5}$$

where $\hat{L}^2(\mathbf{R}^{nd}, n!du^n)$ is the subspace consisting of $L^2(\mathbf{R}^{nd}, n!du^n)-$ functions which is symmetric in n d-dimensional variables.

With the isomorphism (5) we can introduce a Hilbert space $(L^2)^-$ of generalized white noise functionals. Namely,

$$(L^2)^- = \sum_{n=0}^{\infty} \oplus \mathcal{H}_n^-, \tag{6}$$

where $\mathcal{H}_n^- \simeq \hat{H}^{-(nd+1)/2}(\mathbf{R}^{nd})$ (= space of symmetric $H^{-(nd+1)/2}(\mathbf{R}^{nd})-$ functions, denoting by H^m the Sobolev space of order m). The norm $\| \ \|$ in $(L^2)^-$ is defined by $\| \ \|^2 = \sum_n c_n^2 \| \ \|_{-n}^2$ with $\mathcal{H}_n^-$-norm $\| \ \|_{-n}$ and with a decreasing sequence c_n^2 which is chosen suitably depending on the request that how large the $(L^2)^-$ should be.

Theoretically speaking, a space of test functionals should first be introduced, although we have skipped the stage of giving the full definition. The space is denoted by $(L^2)^+$ and there is a Gel'fand triple:

$$(L^2)^+ \subset (L^2) \subset (L^2)^-.$$

It should be noted that there is another useful space of generalized white noise functionals denoted by (S^*), which is defined like the S^* the dual space of the Schwartz space S on $\mathbf{R}^1$. For the present purpose $(L^2)^-$, in particular $\mathcal{H}_1^-$, is more convenient as will be seen in the next section.

Returning to $(L^2)^-$, we then introduce a differential operator

$$\partial_t = \frac{\partial}{\partial x(t)}, \quad t \in \mathbf{R}^d.$$

Since the set $\{x(t) : t \in \mathbf{R}^d\}$ is taken to be the system of variables of $(L^2)^-$ —functionals, it is quite reasonable, in an intuitive level, to consider such an operator as ∂_t to define a partial differential operator. Mathematically rigorous definition of the operator ∂_t is given below:

$$\partial_t \varphi = S^{-1} \frac{\delta}{\delta \xi(t)} S\varphi. \tag{7}$$

Its adjoint operator ∂_t^* can also be defined and it acts on $(L^2)^-$ as a <u>creation</u> operator, while ∂_t itself acts as an <u>annihilation</u> operator, the domain of which is wide enough including $(L^2)^+$. The system of these operators $\{\partial_t, \partial_s^* : t, s \in \mathbf{R}^d\}$ plays very important roles in our causal calculus.

Incidentally, we note that the multiplication operator by $x(t)$ is expressible as

$$x(t)\cdot = \partial_t + \partial_t^*.$$

2. White noise on a manifold

We are interested in the case where the parameter of a Gaussian random field is taken to be a C^∞-manifold in $\mathbf{R}^d$, instead of a point in an Euclidean space. If we wish to express such a random field as a functional of white noise, and/or if we wish to discuss variational calculus, we have to consider a white noise on a domain in $\mathbf{R}^d$ or on a lower dimensional manifold. Since such a white noise is not an ordinary random field, it is necessary to provide a technique how to restrict the parameter of a random field to a lower dimensional subset of $\mathbf{R}^d$.

Let C be a $(d-1)$-dimensional C^∞-manifold which is a convex subset of $\mathbf{R}^d$ and is diffeomorphic to the sphere S^{d-1}. The collection of such C's is denoted by $\mathcal{C}$. The symbol $d\sigma$ stands for the surface element of C in $\mathcal{C}$. Starting from $L^2(C, d\sigma)$, we form a Gel'fand triple, and a white noise measure μ^C can be introduced in the space of generalized functions on C. To fix the idea we take the following triple:

$$H^\alpha(C) \subset L^2(C, d\sigma) \subset H^{-\alpha}(C), \qquad \alpha > \frac{d}{2}.$$

The white noise measure μ^C is uniquely determined by the characteristic functional

$$C(\xi) = \exp[-\frac{1}{2}\|\xi\|_C^2], \qquad \| \; \|_C : \text{ the } L^2(C, d\sigma) - \text{norm}.$$

A member x^C of $H^{-\alpha}(C)$ is a sample function of the white noise with parameter set C.

Now one may ask if there is any nice natural relation between μ and μ^C. Before answering to this question, we remind some properties of a generalized function. A member x^C of $H^{-\alpha}(C)$ is viewed as a "restriction" of an element, say x, of E^* to the manifold C. More precisely, the x is understood as a generalized function (a Schwartz distribution) on $\mathbf{R}^d$ with singular support C. This fact can be observed by taking a test functon as follows. Take a function f from the Sobolev space $H^{\alpha+1/2}(\mathbf{R}^d)$ to have the bilinear form

$$\int_C x^C(u(s))f(u(s))d\sigma(s) = < x^C, f^C >, \tag{8}$$

where f^C is the restriction of f to C. By the trace theorem for the Sobolev space we know that f^C belongs to $H^\alpha(C)$, so that the above bilinear form makes sense. On the other hand, the formula (8) is a continuous linear functional of f, which means that there is a generalized function x such that we can let it be supported by C ; namely we may write

$$< x^C, f^C > = \int x(u)\delta_C(u)f(u)du^d, \tag{9}$$

where $\delta_C(u)$ is the delta-function with the support C. This enables us to understand that we have been given a member in E^* which may be formally denoted by $x \cdot \delta_C$.

In short, such a formal relatonship can clearly be explained by applying the S-transform. Indeed, both (8) and (9) are carried by S to the same expression

$$\int f(u)\delta_C(u)\xi(u)du^d = < f^C, \xi^C >, \quad \xi \in E,$$

which has two aspects; one is the S-transform of a generalized white noise functional on E^*, and the other is that of an ordinary white noise functional with the parameter restricted to C. The possibility to get such a unified expression is, in fact, a big advantage of the use of the integral representaton for generalized white noise functionals in terms of the S-transform.

The next method of restriction the parameter is the use of the creation operators ∂_t^*. Note that

$$S\partial_t^* = \xi(t) \quad \text{(multiplication by } \xi(t)) \tag{10}$$

We therefore see that setting $\partial^*(f) = \int_{R^d} f(u)\partial_u^* du^d$ with $f \in E$

$$S\{\partial^*(f)1\}(\xi) = (f, \xi), \quad (\ ,\ ) : \text{inner product in } L^2(\mathbf{R}^d), \tag{11}$$

holds. The formula (11) shows that f can be taken to be a generalized function. Let the singular support of f be the manifold C. Then the inner product in the formula (11) turns out to be the bilinear form $< f, \xi^C >$ with ξ^C the restriction of ξ to C. If ξ is in $H^{\alpha+1/2}(\mathbf{R}^d)$, then ξ^C is a member of $H^\alpha(C)$, and hence the bilinear form makes sense even in the case where the restriction of f to C is in

$H^{-\alpha}(C)$. We can then play the same game as in the case where the formulas (8) and (9) are concerned.

Through the above observations, we can see how two white noises on $\mathbf{R}^d$ and on C are connected and eventually see how μ and μ^C are related. Further observation related to conditional distribution will be discussed in a separate paper.

3. Innovation approach

We are now ready to discuss an innovation theory for a Gaussian random field indexed by a manifold, where variational calculus can be applied. To fix the idea we shall restrict our attention to the two dimensional parameter case (i.e. $d = 2$) and discuss two typical but rather simple examples of a Gaussian random fields which are depending on a curve C .

The class $\mathcal{C}$ in which a curve C runs through is defined as follows:

$$\mathcal{C} = \{C : C^\infty \text{ convex plane curve diffeomorphic to } S^1\}.$$

Let $0(E)$ be the rotation group of E (see e.g. [3]), where E is now taken to be the $D_0 = \{\xi \in C^\infty : w\xi \in C^\infty\}$ with $w\xi(u) = \xi(u/|u|^2)|u|^{-2}$, and define

$$G(\mathcal{C}) = \{\tilde{g} : \text{diffeomorphism of } \overline{\mathbf{R}} \equiv \mathbf{R} \cup \{\infty\}, \xi(\tilde{g}\, u)|J(\tilde{g}\, u)|^{1/2} \in 0(E)\},$$

where J means the Jacobian. To get the variation of functionals on $\mathcal{C}$, the curves in $\mathcal{C}$ are assumed to be deformed only be the actions of the group $G(\mathcal{C})$.

The first random field to be discussed is given by

$$X(C) = \int\int_{[C]} F(C,u)x(u)du^2, \tag{11}$$

where $[C]$ denotes a closed domain with boundary C and where x is a sample function (generalized function) of a white noise. Apply the S-transform to obtain a U-functional of the form

$$U(C)(\xi) = \int\int_{[C]} F(C,u)\xi(u)du^2, \tag{12}$$

where ξ in this expression is taken to be an element of $H^{3/2}(\mathbf{R}^2)$.

We are now ready to obtain the variation when C varies within the class $\mathcal{C}$. For this purpose, we assume that the kernel $F(C,u)$ is in $H^{3/2}(\mathbf{R}^2)$ as a function of u and that the variation $\delta F(C,u)$ have the functional derivative $F'_n(C,u)(s)$ square intergable in u.

Theorem 1. *Under the assumption on $F(C,u)$ mentioned above we have the variation expressed in the form*

$$\begin{aligned} \delta X(C) = &\int\int_{[C]}\int F'_n(C,u)(s)\delta n(s)x(u)dsdu^2 \\ &+ \int_C F(C,u(s))x(u(s))\delta n(s)ds, \quad ds : \textit{line element}, \end{aligned} \tag{13}$$

when C changes to $C + \delta C$ which differs from C by δn in the normal direction.

Proof. Apply the S-transform to $X(C)$ to obtain the formula (12) and take the variation $\delta U(\xi)$. Then we have

$$\begin{aligned}(14)\qquad \delta U(C)(\xi) &= \int\int_{[C]} \delta F(C,u)\xi(u)du^2 + \int_C F(C,u(s))\xi(u(s))\delta n(s)ds \\ &= \int\int_{[C]}\int_C F'_n(C,u)(s)\delta n(s)\xi(u)dsdu^2 \\ &\quad + \int_C F(C,u(s))\xi(u(s))\delta n(s)ds.\end{aligned}$$

The last term of the above equation can be understood as in the last section, and it is considered as the $S-$transform of a generalized white noise functional which is permitted to write in the form of the second term in the formula (13), following the interpretation given in the last section. □

There is an important remark. The first and the second terms can be discriminated as is shown in [11] by K.-S. Lee. Indeed, if δn is small, then the first term of (13) is an infinitesimal random variable of order $\| \delta n \|$, the $L^2(C, ds)-$norm of δn, while the second term has a variance which is infinite multiple of $\| \delta n \|$, since the $L^2(\mathbf{R}^2)-$norm of $\delta-$function is involved.

We now restrict our attention to the second term of (13) which is more interesting for us since it appears in many applications (e.g. see [12]). The technique of restricting the parameter enables us to regard the second term as a random variable formed by the white noise on C. It can therefore be written as

$$\int_C F(C,u(s))x^C(s)\delta n(s)ds.$$

The class $\mathcal{C}$ has enough curves to let δn moves in a dense subset of $L^2(C, ds)$ under the action of the group $G(\mathcal{C})$. If it is assumed that $F(C, u(s)) \neq 0$ on C, then we can get the white noise x^C, and hence $\{x^C : C \in \mathcal{C}\}$ can be regarded as the *innovation* of the given Gaussian random field $\{X(C) : C \in \mathcal{C}\}$.

The "canonical property" can also be discussed under the present setting as has been done in [8].

The next class which interests us involves random fields depending on a curve C that is expressed in the form

$$(15)\qquad Y(C) = \int_C F(u(s))x(u(s))ds, \qquad C \in \mathcal{C}.$$

To avoid unnecessary complexity we assume that the kernel $F(u)$ is independent of C, but it is assumed to be smooth enough.

Such a field appears in many cases; a good example comes from the above theorem, where we are given a white noise integral over a boundary curve, and another interesting example can be seen where we discuss the conditional expectation given the values of the field on the curve C (see [12] and [13] examples for Lévy's Brownian motion).

Retuning to the field $\{Y(C) : C \in \mathcal{C}\}$ given by (15), we prove the following theorem.

Theorem 2. *The variation of $Y(C)$ is expressed in the form*

$$\delta X(C) = \int_C [\frac{\partial}{\partial n} F(u(s)) \cdot x(u(s)) + F(u(s)) x'_n(s) - \kappa F(u(s) x(u(s))] \delta n(s) ds, \tag{16}$$

where κ is the curvature of the curve.

Proof. The S-transform of $Y(C)$ is

$$U(C, \xi) = \int_C F(u(s)) \xi(u(s)) ds, \tag{17}$$

where ξ may be taken from the space $H^{3/2}(\mathbf{R}^2)$. Hence the restriction $\xi(u(s))$ with $u(s) \in C$ of $\xi(u)$ to the curve C is a member of the Sobolev space $H^1(C)$. With this remark we are now ready to take the variation of $U(C, \xi)$:

$$\delta U(C, \xi) = \int_C [\frac{\partial}{\partial n} F(u(s)) \cdot \xi(u(s)) + F(u(s)) \frac{\partial}{\partial n} \xi(u(s)) - \kappa F(u(s)) \xi(u(s))] \delta n(s) ds. \tag{18}$$

This equation can be rephrased in terms of functionals of x. There is no difficulty to do so for the first and the third terms in the integrand [ ]. For the second term we should note that $\frac{\partial}{\partial n} \xi$ can be associated with the normal derivative $\frac{\partial}{\partial n} X \equiv X'_n$ which is a generalized function, and we can restrict it to C in a same manner as we did in the last section. We therefore come to the conclusion. □

With the variational calculus for Gaussian random fields that we have discussed so far, we shall be able to develop white noise analysis from the viewpoint of random functional analysis when a curve (or more generally, a manifold) is taken to be a parameter.

Remark 1. Similar arguments as before, we can get the innovation of the $Y(C)$.

Remark 2. Another important remark is that the deformation of C can be limited within the action by the conformal transformations that has been discussed in [11]. There the unitary representation theory of Lie groups has close connection with the canonical property through the variational calculus.

References

[1] P. Lévy, *Problèmes concrets d'analyse fonctionnelle.* Gauthier-Villars, 1951.

[2] –, A special problem of Brownian motion, and a general theory of Gaussian random functions. *Proc. 3rd Berkeley Symp. on Math. Statistics and Probability,* II (1956), ed, J. Neyman, 133-175, §3 and §6.

[3] T. Hida, *Brownian motion.* Iwanami (in Japanese), 1975; English ed. Springer-Verlag, 1980.

[4] –, *Analysis of Brownian functionals,* Carleton Math. Lecture Notes no. 13. Carleton Univ. 1975.

[5] –, White noise analysis and Gaussian random fields. Stochastic methods in Mathematics and Physics. *Proc. 24th Karpacz Winter School on Theoretical Physics. ed. R. Gielerak and W. Karwowski,* World Scientific. 1989, 277-289.

[6] T. Hida, H.-H. Kuo, J. Potthoff and L. Streit, *White noise analysis,* World Scientific, 1990.

[7] T. Hida, K.-S. Lee ans S.-S. Lee, Conformal invariance of white noise. *Nagoya Math. Journal,* 98 (1985), 87-98.

[8] T. Hida, K.-S. Lee and Si Si, Multi-dimensional parameter white noise and Gaussian random fields. *Recent advances in communication and control theory.* ed. R.E. Kalman et al. (1987), 177-183.

[9] T. Hida and K. Saito, *Introduction to white noise analysis.* BiBoS Notes 400/89, Universität Bielefeld, 1989, 5-23.

[10] T. Hida and Si Si, Variational calculus for Gaussian random fields. *Proc. IFIP Warsaw Conference* 1988. ed. J. Zabczyk, to appear.

[11] K.-S. Lee, White noise approach to Gaussian random fields, to appear.

[12] Si Si, A note on Lévy's Brownian motion. *Nagoya Math. Journal,* 108 (1987), 121-130.
The same title, II, *Nagoya Math. Journal,* 114 (1989), 165-172.

[13] –, *Gaussian processes and conditional expectations.* BiBoS Notes 292/87, Universität Bielefeld. 1987.

Connectivity of certain graphs on halfspaces, quarter spaces, ...*

Harry Kesten

Abstract. On the vertex set $\mathcal{V} = \mathcal{V}(d,e) := \mathbb{Z}^d \times \mathbb{Z}_+^e$ connect u and v by an edge with probability $p(v-u)$. Assume all possible edges are independent. Denote the resulting random graph by $\mathcal{G} = \mathcal{G}(d,e)$. Under mild symmetry and irreducibility assumptions on the function p we prove that $P\{\mathcal{G}$ is connected$\} = 0$ (respectively 1) if $\sum_{u \in \mathbb{Z}^{d+e}} p(u) < \infty$ (respectively $\sum_{u \in \mathbb{Z}^{d+e}} p(u) = \infty$).

Keywords and Phrases. Random graphs, connectivity.

1. Introduction and statement of result

Grimmett, Keane and Marstrand [GKM] considered connectedness of the random graph $\mathcal{G}(d,0)$ with vertex set $\mathbb{Z}^d$, and an edge between u and v present with probability $p(v-u)$, for some given function $p : \mathbb{Z}^d \to [0,1)$ which satisfies $p(u) = p(-u)$. (With the exception of Remark (iii) below, only undirected graphs are considered here; for any unordered pair of vertices u, v with $u \neq v$ there is either one edge between them in $\mathcal{G}$ or none.) Under an obvious irreducibility assumption (see (1.5) below) it is shown in [GKM] that if all edges are selected independently of each other, then

$$(1.1) \qquad P\{\mathcal{G}(d,0) \text{ is connected}\} = 0 \text{ if } \sum_{u \in \mathbb{Z}^d} p(u) < \infty$$

and

$$(1.2) \qquad P\{\mathcal{G}(d,0) \text{ is connected}\} = 1 \text{ if } \sum_{u \in \mathbb{Z}^d} p(u) = \infty.$$

(1.1) is easy, since $\sum p(u) < \infty$ implies for each vertex v

$$P\{v \text{ is isolated}\} = \prod_u [1 - p(u-v)] = \prod_u [1 - p(u)] > 0.$$

Thus, the ergodic theorem implies that there are w.p.1 infinitely many isolated vertices in this case.

(1.2) is considerably more difficult. The proof in [GKM] breaks down if the vertex set $\mathbb{Z}^d$ is replaced by $\mathbb{Z}^d \times \mathbb{Z}_+^e$ for $e \geq 1$ ($\mathbb{Z}_+ = \{0,1,2,...\}$) because the

*Research supported by the NSF through a grant to Cornell University.

Probability Theory

latter is no longer translation invariant. A simpler proof of (1.2) was given in [AKN], and a still simpler one in [GKN]. But these proofs too fail for $\mathbb{Z}^d \times \mathbb{Z}^e_+$. Kalikow and Weiss [KW] gave a proof of (1.2) for $d = 0$, $e = 1$. Thus, [KW] handles $\mathcal{G}$(0,1) which has the "halfline" $\{0, 1, 2, ...\}$ as vertex set. A different proof for $\mathcal{G}$(0,1) is given in [GKN]. The proof of Kalikow and Weiss is the starting point for this paper. We use it to prove by induction the analogues of (1.1) and (1.2) for "half spaces" ($\mathcal{G}$(d,1)), "quarter spaces" ($\mathcal{G}$(d,2)) etc. The precise result is as follows.

Theorem. *Let $p : \mathbb{Z}^{d+e} \to [0,1)$ be invariant under changes of all signs simultaneously, or of the last e coordinates separately, i.e.,*

$$p(u) = p(-u) \text{ and} \tag{1.3}$$

$$p(u_1, ..., u_d, \pm u_{d+1}, ..., \pm u_{d+e}) = p(u) \text{ for all} \tag{1.4}$$

$$2^e \text{ choices of the signs } + \text{ and } - .$$

Assume further that p satisfies the following irreducibility condition

$$\text{for each } u \in \mathcal{V}(d,e) \tag{1.5}$$

$$\text{there exists a finite sequence } (u^{(1)}, ..., u^{(k)}) \subset \mathbb{Z}^{d+e}$$

$$\text{such that } p(u^{(i)}) > 0, u^{(1)} + ... + u^{(i)} \in \mathcal{V}(d,e), 1 \le i \le k,$$

$$\text{and } u^{(1)} + ... + u^{(k)} = u.$$

Then

$$P\{\mathcal{G}(d,e) \text{ is connected}\} = 0 \text{ if } \sum_{u \in \mathbb{Z}^{d+e}} p(u) < \infty, \tag{1.6}$$

and

$$P\{\mathcal{G}(d,e) \text{ is connected}\} = 1 \text{ if } \sum_{u \in \mathbb{Z}^{d+e}} p(u) = \infty. \tag{1.7}$$

Remarks. (i) Note that we required $p(u) < 1$ for all u. If $p(u_0) = 1$ for some $u_0 \neq 0$, then one easily obtains under (1.3)–(1.5) that $\mathcal{G}(d,e)$ is connected w.p. 1.

(ii) The irreducibility condition (1.5) obviously cannot be dispensed with. It is, however, sufficient to make the requirement (1.5) only for $u = \pm\xi_i, 1 \le i \le d$, and $u = \xi_i$, $d+1 \le i \le d+e$, where $\xi_i = (0,0,...,1,0,...0)$ with the 1 in the i-th place. Indeed if (1.5) holds for some u' and u'', then it also holds for $u' + u''$.

(iii) The symmetry assumption (1.4) seems stronger than necessary, but we have been unable to prove (1.7) without this condition. Simple monotonicity considerations show that we may drop (1.4) if we strengthen the condition in (1.7) to

$\sum \underline{p}(u) = \infty$, where $\underline{p}(u) := \min p(u_1, ..., u_d, \pm u_{d+1}, ..., \pm u_{d+e})$ with the minimum taken over all possible combinations of the signs.

It is also possible to prove (1.7) for $e = 1$ without (1.4), by a variation of the proof in the next section. If one drops (1.4) one should probably strengthen the hypothesis $\sum p(u) = \infty$ to

$$\sum_{\substack{u_j \in \mathbb{Z}, i \leq j \leq d \\ \epsilon_i u_i \geq 0, d+1 \leq i \leq d+e}} p(u_1, ..., u_{d+e}) = \infty, \tag{1.8}$$

for every choice of $\epsilon_i \in \{-1, +1\}, d+1 \leq i \leq d+e$. (Here u_i is the i-th component of u.) It seems that in order to prove that (1.3), (1.5) and (1.8) imply $P\{\mathcal{G}(d,e)$ is connected$\} = 1$, one should consider the following "directed version" of our problem. Let $\hat{\mathcal{G}}(d,e)$ be the *directed* graph with vertex set $\mathcal{V}(d,e)$ which contains a directed edge from u to v with probability $p(v-u)$. Assume all directed edges are independent. (Thus there are now two independent choices for edges between the given vertices u and v: one from u to v and one from v to u.) We no longer assume (1.3).

Problem. Is it true that (1.8) and a suitable irreducibility condition imply

$$P\{\text{For all } u, v \in \mathcal{V}(d,e), u \neq v, \text{ there is a directed path in } \mathcal{V}(d,e) \text{ from } u \text{ to } v\} = 1? \tag{1.9}$$

It seems likely that if one can prove (1.9) for the halfline ($d = 0$, $e = 1$), that then an analogue of our inductive proof can be used to handle general d and e.

(iv) If $\sum p(u) = \infty$, then w.p. 1 each vertex is incident to infinitely many edges. Thus connectedness of $\mathcal{G}(d,e)$ is equivalent to $\mathcal{G}(d,e)$ having at most one infinite component in this case. This observation was the basis of the proof at (1.2) in [AKN]. This led Gandolfi (private communication) to investigate the more general question whether $\mathcal{G}(d,e)$ always has at most one infinite component (under (1.3)–(1.5)), even if $\sum p(u) < \infty$. Very recently Grimmett and Marstrand [GM] answered this question affirmatively in the finite range case (i.e., when there exists some finite k such that $p(u) = 0$ for $|u| > k$); see also [BGN] for partial results. The general infinite range case still seems to be open, though.

Acknowledgement. The author thanks A. Gandolfi, G. Grimmett and C.M. Newman for several helpful conversations.

2. Proof of the theorem

It will be necessary to work with several probability distributions on the edges. To distinguish the various distributions we shall write P_r for the probability measure which governs $\mathcal{G}$ when the edge between u and v is included with probability $r(v -$

u). In addition we use the following notation. $\{u,v\}$ denotes the (potential) edge between u and v. $E(u,v)$ is the event that $\{u,v\}$ is present in $\mathcal{G}$. $u \leftrightarrow v$ means that there exists a sequence of vertices $u_1, u_2, ..., u_k$ such that, with $u_0 = u$, $u_{k+1} = v$, each $\{u_i, u_{i+1}\}$, $0 \le i \le k$, is present. We write $0(d+e)$ for the origin in $(d+e)$ dimensions; this is a vertex in $\mathcal{V}(d,e)$. For $u \in \mathcal{V}(d,e), H(u)$ will be the halfline

$$H(u) = \{u\} \times \mathbb{Z}_+ \subset \mathbb{Z}^d \times \mathbb{Z}_+^{e+1},$$

and

$$H = H(0(d+e)).$$

The following lemma contains our fundamental induction step.

Lemma. *Assume that for some $d, e \ge 0, d+e > 0$, (1.7) holds for all p satisfying (1.3)–(1.5). Then, for any $r : \mathbb{Z}^{d+e+1} \to [0,1)$ which is invariant under sign changes and irreducible (i.e., which satisfies the analogues of (1.3)-(1.5) with e replaced by $(e+1)$ one has*

(2.1) $$P_r\{0(d+e+1) \text{ is connected to infinitely many points of } H \text{ in } \mathcal{G}(d,e+1)\} = 1.$$

Proof. We say that we *examine* the edge $\{v,w\}$ if we decide whether $E(v,w)$ occurs or not; in other words, if we fix the value of the indicator function $I_{E(v,w)}$. In this proof we shall examine the edges in various random sets $\mathcal{E} \subset \mathcal{V}(d,e+1) \times \mathcal{V}(d,e+1)$. These $\mathcal{E}$ will be *self determined*, in the sense that for any fixed set $\bar{\mathcal{E}}$, the event $\{\mathcal{E} = \bar{\mathcal{E}}\}$ is determined only by the $E(v,w)$ with $\{v,w\} \in \bar{\mathcal{E}}$. Then by the independence of all edges we still have for any finite set of edges $\{v_1, w_1\}, ..., \{v_k, w_k\}$ that

$$P_r\{E(v_i, w_i), 1 \le i \le k | \mathcal{E}, I_{E(v,w)} \text{ for } \{v,w\} \in \mathcal{E}\}$$
$$= \prod_{i=1}^{k} r(w_i - v_i) \text{ on } \{\{v_i, w_i\} \notin \mathcal{E} \text{ for } 1 \le i \le k\}.$$

In other words, when $\mathcal{E}$ and $I_{E(v,w)}$ for $\{v,w\} \in \mathcal{E}$ are given, the edges outside $\mathcal{E}$ are conditionally independent and governed by the original probabilities.

The basis of our proof is a coupling of $\mathcal{G}(d,e)$ with a random graph $\mathcal{H}$ whose vertex set is a subset of $\mathcal{V}(d,e+1)$. Here $\mathcal{G}(d,e+1)$ is governed by P_r, but $\mathcal{G}(d,e)$ will be governed by P_p for the p of (2.3) below. The coupling will be such that

(2.2) if $0(d+e) \leftrightarrow u$ in $\mathcal{G}(d,e)$, then also $0(d+e+1) \leftrightarrow H(u)$ in $\mathcal{H}$.

From the assumption (1.7) it will follow that w.p.1 $0(d+e) \leftrightarrow u$ in $\mathcal{G}(d,e)$ for every $u \in \mathcal{V}(d,e)$. Together with (2.2) this will imply that $0(d+e+1) \leftrightarrow H(u)$ in $\mathcal{H}$ for every $u \in \mathbb{Z}^d \times \mathbb{Z}_+^e$. Lastly, in constructing $\mathcal{H}$ we shall only examine edges in some set $\mathcal{E}$. The edges outside $\mathcal{E}$ can still be examined and these together with the edges of $\mathcal{H}$ will give infinitely many connections between $0(d+e+1)$ and H in $\mathcal{G}(d,e+1)$.

We turn to the details. In the remainder of the proof, u (respectively v, w) will stand for a generic vertex of $\mathcal{V}(d,e)$ or $\mathbb{Z}^{d+e}$ (respectively $\mathcal{V}(d,e+1)$ or $\mathbb{Z}^{d+e+1}$). We define for $u \in \mathcal{V}(d,e)$ (i.e. for $u_{d+1} \geq 0, \dots, u_{d+e} \geq 0$)

(2.3) $$p(u) = P_r\{\text{there exists an edge in } \mathcal{G}(d,e+1) \text{ between } 0(d+e+1) \text{ and some vertex on } H(u)\} = 1 - \prod_{v \in H(u)} (1 - r(v)).$$

We extend p to all of $\mathbb{Z}^{d+e}$ by setting

$$p(u_1, \dots, u_d, \pm u_{d+1}, \dots, \pm u_{d+e}) = p(u)$$

for all choices of the 2^e signs. Then p satisfies (1.4). It also satisfies (1.3), because $r(u) = r(-u)$. One also easily deduces (1.5) for p from its analogue for r. Finally we note that our extension of p is such that for all $u \in \mathcal{V}(d,e)$, $v \in \mathcal{V}(d,e+1)$

(2.4) $$P_r\{\text{there exists an edge in } \mathcal{G}(d,e+1) \text{ between } v \text{ and } \{(u_1, \dots, u_{d+e}, z) : z \geq v_{d+e+1}\}\} = p(u - v') = 1 - \prod_{w \in H(u-v')} (1 - r(w)),$$

where $v' = (v_1, \dots, v_{d+e})$ is the projection of v onto $\mathcal{V}(d,e)$.

We now break the proof up into two cases:

(α) $$\sum_{u \in \mathbb{Z}^{d+e}} p(u) < \infty,$$

(β) $$\sum_{u \in \mathbb{Z}^{d+e}} p(u) = \infty.$$

Case (α). In this case $0(d+e+1)$ is connected in $\mathcal{G}(d,e+1)$ to only finitely many $H(u)$ with P_r-probability 1 (by the Borel-Cantelli lemma). On the other hand, there are with P_r-probability 1 infinitely many edges incident to $0(d+e+1)$ in $\mathcal{G}(d,e+1)$. Thus, w.p.1 there exists a (random) $u \in \mathcal{V}(d,e)$ such that there are infinitely many edges between $0(d+e+1)$ and $H(u)$ in $\mathcal{G}(d,e+1)$. Thus there exists a nonrandom $u_0 \in \mathcal{V}(d,e)$ such that

$$P_r\{\text{there exist infinitely many edges between } 0(d+e+1) \text{ and } H(u_0)\} > 0,$$

and again by the Borel-Cantelli lemma

(2.5) $$\sum_{v \in H(u_0)} r(v) = \infty.$$

(Of course, this can also be obtained analytically.) Now by the symmetry and irreducibility assumptions on r, there exists a sequence $v^{(1)}, \dots, v^{(k)}$ such that

(2.6) $$v^{(1)} + \dots + v^{(k)} \in H(-u_0),\ r(v^{(i)}) > 0,\ 1 \leq i \leq k.$$

In fact, we can even choose the $v^{(i)}$ such that for each $v \in H(u_0)$

$$(2.7) \qquad v + \sum_{j \le i} v^{(j)} \in \mathcal{V}(d, e+1), \ 1 \le i \le k.$$

This is quite trivial if $r(w) > 0$ for all $w \neq 0(d+e+1)$, in which case we can take $k = 1$ and $v^{(1)}$ arbitrary in $H(-u_0)$. The reader who wishes to skip some technical details may therefore safely jump to (2.11). To prove (2.7) in general, choose $w^{(1)}, ..., w^{(\ell)} \in \mathbb{Z}^{d+e+1}$ such that

$$(2.8) \qquad r(w^{(i)}) > 0, \ 1 \le i \le \ell, \ w^{(1)} + ... + w^{(\ell)} \in H(-u_0).$$

Let

$$L = \max_{1 \le i \le \ell} |w^{(1)} + ... + w^{(i)}|.$$

Next find $x^{(1)}, ..., x^{(m)} \in \mathbb{Z}^{d+e+1}$ such that

$$(2.9) \quad r(x^{(i)}) > 0, x^{(1)} + ... + x^{(i)} \in \mathcal{V}(d, e+1), 1 \le i \le m, \text{ and } x^{(1)} + ... + x^{(m)}$$
$$= (L, L, ..., L) \in \mathcal{V}(d, e+1).$$

The w's and x's can be found by virtue of the analogues of (1.3) - (1.5) for r. We claim that the sequence $x^{(1)}, ..., x^{(m)}, w^{(1)}, ..., w^{(\ell)}, -x^{(m)}, ..., -x^{(1)}$ can be taken for $v^{(1)}, ...v^{(k)}$ (with $k = 2m + \ell$). Indeed, (2.6) for this sequence is clear from (2.8) and (2.9). Also, since $v \in H(u_0)$ the last $(e+1)$ components of v are positive so that by (2.9)

$$v + \sum_{j \le i} v^{(j)} = v + \sum_{j \le i} x^{(j)} \in \mathcal{V}(d, e+1)$$

if $i \le m$. For $m < i \le m + \ell$ we have

$$v + \sum_{j \le i} v^{(j)} = v + \sum_{j \le m} x^{(j)} + \sum_{1}^{i-m} w^{(j)} = v + (L, L, ..., L) + \sum_{1}^{i-m} w^{(j)}.$$

This too lies in $\mathcal{V}(d, e+1)$ by our choice of L. Finally, for $m + \ell < i \le 2m + \ell$ we have

$$v + \sum_{j \le i} v^{(j)} = v + \sum_{1}^{\ell} w^{(j)} + \sum_{j \le 2m + \ell - i} x^{(j)},$$

and this lies in $\mathcal{V}(d, e+1)$, by virtue of (2.9) and the fact that

$$(2.10) \qquad v + \sum_{1}^{\ell} w^{(j)} = (0, 0, ..., 0, y) \text{ for some } y \ge 0,$$

which follows from $v \in H(u_0), \sum w^{(j)} \in H(-u_0)$. Thus (2.7) holds for all $i \le 2m + \ell = k$.

Now fix some $v^{(1)},...,v^{(k)}$ satisfying (2.6) and (2.7), and introduce the event

$$(2.11)\quad F(v) = E(0(d+e+1),v) \cap E(v,v+v^{(1)}) \cap E(v+v^{(1)},v+v^{(1)}+v^{(2)}) \cap ... \cap E(v+v^{(1)}+...+v^{(k-1)},v+v^{(1)}+...+v^{(k)}).$$

Then

$$P_r\{F(v)\} \geq r(v)\prod_1^k r(v^{(i)}),$$

so that by (2.5)

$$\sum_{v \in H(u_0)} P_r(F(v)) = \infty.$$

The events $\{F(v) : v \in H(u_0)\}$, are not necessarily independent. But if $M > \max_{i \leq k} |v^{(1)} + ... + v^{(i)}|$, then for any subset $\tilde{H}$ of $H(u_0)$ with the property $|v' - v''| > 2M$ for any $v', v'' \in \tilde{H}, v' \neq v''$, the events $\{F(v) : v \in \tilde{H}\}$ are independent under P_r. Thus, by the Borel-Cantelli lemma we have for some $j \leq 2M+1$

$$P_r\{\text{infinitely many } F(v) \text{ with } v \in H(u_0) \text{ occur}\}$$
$$\geq P_r\{F(u_0 + (j + n(2M+1))\xi_{e+d+1} \text{ occurs for infinitely many } n\} = 1.$$

This proves the lemma in case (α), since $F(v)$ implies $0 \leftrightarrow v + v^{(1)} + ... + v^{(k)} \in H$ in $\mathcal{G}(d, e+1)$ for $v \in H(u_0)$.

Case (β). It is only in this case that we use the coupling outlined in the beginning of the proof. We successively define two increasing sequences of graphs $\mathcal{G}_n$ and $\mathcal{H}_n$, with vertex sets $\mathcal{V}_n \subset \mathcal{V}(d,e)$ and $\mathcal{W}_n \subset \mathcal{V}(d,e+1)$, respectively. In some sense $\mathcal{G}_n$ is the projection of $\mathcal{H}_n$ in the direction of the last coordinate axis on $\mathbb{Z}^d \times \mathbb{Z}_+^e \times \{0\}$, which is identified with $\mathcal{V}(d,e)$. $\mathcal{G}_0$ and $\mathcal{V}_0$ consist only of the vertex $0(d+e)$. Similarly $\mathcal{H}_0$ and $\mathcal{W}_0$ consist only of $0(d+e+1)$. Assume now that $\mathcal{G}_n$ and $\mathcal{H}_n$ have already been found in such a way that

$$(2.12)\quad \mathcal{W}_i \cap H(u) \neq \phi \text{ if and only if } u \in \mathcal{V}_i,\ 1 \leq i \leq n.$$

To construct $\mathcal{G}_{n+1}$ and $\mathcal{H}_{n+1}$ define for any $u \in \mathcal{V}_n, v(u)$ as the vertex in $\mathcal{W}_n \cap H(u)$ with the minimal last coordinate. This vertex is of the form

$$v(u) = (u_1, ..., u_{d+e}, z(u)) \text{ for some } z(u) \geq 0.$$

Now we examine edges between any of the $v(u), u \in \mathcal{V}_n \setminus \mathcal{V}_{n-1}$ and

$$(2.13)\quad w \in \mathcal{V}(d,e+1) \setminus \bigcup_{u \in \mathcal{V}_n} H(u) \text{ with } w_{d+e+1} \geq z(u)$$

for inclusion in $\mathcal{H}_{n+1}$. The edge $\{v(u), w\}$ (for $u \in \mathcal{V}_n \setminus \mathcal{V}_{n-1}$ and w satisfying (2.13)) is included in $\mathcal{H}_{n+1}$ with probability $r(w - v(u))$. All these edges are chosen independently of each other and of all previous choices. Of course if $\{v(u), w\}$ is included in $\mathcal{H}_{n+1}$, then w is to be included in the vertex set $\mathcal{W}_{n+1}$ of $\mathcal{H}_{n+1}$. $\mathcal{H}_{n+1}$ also includes all of $\mathcal{H}_n$. Finally, for

(2.14) $$u \in \mathcal{V}_n \text{ and } u' \in \mathcal{V}(d,e)\backslash \mathcal{V}_n$$

the edge $\{u, u'\}$ will be included in $\mathcal{G}_n$, if and only if an edge from $v(u)$ to some vertex in $H(u')$ was added to $\mathcal{H}_{n+1}$ in the last stage. In this case u' is of course also included in $\mathcal{V}_{n+1}$. Again $\mathcal{G}_{n+1}$ includes all of $\mathcal{G}_n$. No other edges or vertices then the ones just described are included in $\mathcal{H}_{n+1}$ and $\mathcal{G}_{n+1}$, respectively.

It is clear from the construction that $\{\mathcal{H}_n\}$ and $\{\mathcal{G}_n\}$ are increasing sequences and that (2.12) holds for all n. We also note that at the $(n+1)$-th stage, when $\mathcal{H}_{n+1}$ is constructed, we only examine edges with one endpoint a $v(u) \in \mathcal{W}_n \backslash \mathcal{W}_{n-1}$, and the other endpoint w satisfying (2.13). Since neither of these endpoints lies in $\mathcal{W}_0 \cup \ldots \cup \mathcal{W}_{n-1} = \mathcal{W}_{n-1}$, we see that these edges have not been examined before in the construction of $\mathcal{H}_1, \ldots, \mathcal{H}_n$. Moreover, each of the examined edges is included with the same probability as in $\mathcal{G}(d, e+1)$ (under P_r). If we denote $\lim \mathcal{H}_n$ by $\mathcal{H}$ and the collection of all examined edges of $\mathcal{V}(d, e+1) \times \mathcal{V}(e, d+1)$ by $\mathcal{E}$, then $\mathcal{E}$ is self determined and we may view $\mathcal{H}$ as a random subgraph of $\mathcal{G}(d, e+1)$. That is, if, after the construction of $\mathcal{H}$, one adds to $\mathcal{H}$ each edge $\{w', w''\}$ which has not yet been examined with probability $r(w'' - w')$, and one does this independently for all $\{w', w''\} \notin \mathcal{E}$, then one obtains a random graph which has the same distribution (under P_r) as $\mathcal{G}(d, e+1)$.

Next we show that $\lim \mathcal{G}_n$ is stochastically smaller than $\mathcal{G}(d, e)$ under the distribution P_p, but $\cup \mathcal{V}_n$ has the same distribution as the set

$$\{u \in \mathcal{V}(d, e) : 0 \leftrightarrow u \text{ in } \mathcal{G}(d, e)\},$$

again with the distribution P_p used to construct the last set. More precisely we shall show that

(2.15) the distribution of the $\mathcal{V}_n$ constructed above is the same as the distribution under P_p of $\{u \in \mathcal{V}(d, e)$: there exists a path of n edges from 0 to u in $\mathcal{G}(d, e)\}$.

In a way $\lim \mathcal{G}_n$ is a spanning tree for $\mathcal{G}(d, e)$.

To verify this claim we first examine $\mathcal{G}_1$ and $\mathcal{V}_1$. The only edges included in $\mathcal{G}_1$ are edges $\{0(d+e), u\}$ for which an edge $\{0(d+e+1), w\}$ with $w_{d+e+1} \geq 0$ was included in $\mathcal{H}_1$ (see (2.13) for $n = 0$). Thus

$$\begin{aligned} P\{\{0, u\} \in \mathcal{G}_1\} &= P\{\exists \text{ edge from } 0(d+e+1) \text{ to some } w \in H(u) \text{ in } \mathcal{H}_1\} \\ &= 1 - \prod_{w \in H(u)} (1 - r(w)) = p(u) \\ &= P_p\{\{0, u\} \in \mathcal{G}(d, e)\}. \end{aligned}$$

Moreover, all the edges $\{0, u\}$ are independent of each other (since the inclusion of $\{0, u\}$ depends only on edges with one endpoint on $H(u)$, and for distinct u's the corresponding $H(u)$'s are disjoint). Thus (2.15) holds for $n = 1$. By essentially the same argument one proves (by induction on n) that conditionally on $\mathcal{H}_n$, edges $\{u', u''\}$ with $u' \in \mathcal{V}_n \setminus \mathcal{V}_{n-1}$, $u'' \in \mathcal{V}(d,e) \setminus \mathcal{V}_n$, are added to $\mathcal{G}_n$ independently of each other, and with the same probability as they would be included $\mathcal{G}(d,e)$ under P_p. Note for this last step that in constructing $\mathcal{H}_{n+1}$ from $\mathcal{H}_n$ we only examine edges between $\mathcal{V}_n \setminus \mathcal{V}_{n-1}$ and $\mathcal{V}(d,e) \setminus \mathcal{V}_n$, and that those edges have not been examined for the construction of $\mathcal{H}_n$. Also, when examining edges from $u \subset \mathcal{V}_n \setminus \mathcal{V}_{n-1}$ to $\mathcal{V}(d,e) \setminus \mathcal{V}_n$ we only check edges incident to the single point $v(u)$, and (2.4) shows that such edges are added with the same probability as under P_p. Now that we know the conditional distribution of the edges to be added to $\mathcal{G}_n$ at the $(n+1)$-th stage, it is also clear that (2.15) for n implies that (2.15) also holds for $(n+1)$. Thus (2.15) holds for all n.

(2.15) implies

$$P\{\bigcup_{n \geq 0} \mathcal{V}_n = \mathcal{V}(d,e)\} = P_p\{0 \leftrightarrow u \text{ for every } u \text{ in } \mathcal{G}(d,e)\} = 1,$$

where the last equality is true because we assumed (1.7) for our d, e. But then, by the property (2.12) of our construction, we have also

$$P_r\{\text{for each } u \in \mathcal{V}(d,e), \mathcal{G}(d,e+1) \text{ contains a vertex } v \in H(u)\} = 1. \tag{2.16}$$

(2.16), or rather (2.12), verifies (2.2) for our construction.

To complete the proof of the lemma we now first make the extra assumption

$$\sum_{\substack{v \in \mathbb{Z}^{d+e+1} \\ v_{d+e+1} > 0}} r(v) = \infty. \tag{2.17}$$

Observe that in constructing $\mathcal{H}$ we examined edges with one endpoint on H only at the first stage, because $H = H(0(d+e))$ and $0(d+e) \in \mathcal{V}_0$. Since at the first stage $\mathcal{V}_0 = \{0(d+e)\}$ and $v(0(d+e)) = 0(d+e+1)$ all the edges examined at the first stage have one endpoint at the origin. Thus none of the edges with one endpoint in

$$\hat{H} := 0(d+e) \times \{1,2,3,...\}$$

have been examined. Now let $\mathcal{H}$, and the set $\mathcal{E}$ of examined edges be fixed. For each $u \in \mathcal{V}(d,e)$, let $w(u)$ be a vertex in $\mathcal{G}(d,e+1) \cap H(u)$. By (2.16) there exists such a $w(u)$ for each u with P_r-probability 1. Finally, we examine the edges from $w(u)$ to

$$\begin{aligned} 0(d+e) \times \{w(u)_{d+e+1} + 1, w(u)_{d+e+1} + 2, ...\} \\ = (0,0,...,0,w(u)_{d+e+1}) + \hat{H} \subset \hat{H}. \end{aligned}$$

The edge from $w(u)$ to $(0,0,...,0,w(u)_{e+d+1}+z)$ will be added to H with probability

$$r((0,0,...,0,w(u)_{e+d+1} + z) - w(u)) = r((-u_1,...,-u_{d+e},z)).$$

Again we take all these edges with $u \in \mathcal{V}(d,e), z \geq 1$, independent. Because these latest edges had not been examined before in the construction of $\mathcal{H}$, the resulting graph of $\mathcal{H}$ plus the new edges plus their endpoints, may still be regarded as a subgraph of $\mathcal{G}(d,e+1)$ under P_r. Since

$$\sum_{\substack{z>0 \\ u\in\mathcal{V}(d,e)}} r((-u_1,\dots,-u_{d+e},z)) = \infty$$

when (2.17) holds (recall (1.4)), there will with probability 1 be infinitely many edges added to $\mathcal{H}$ in this last stage. In particular, there will be at least one $w(u)$ and $z > 0$ such that

$$0 \leftrightarrow w(u) \text{ in } \mathcal{H} \text{ and the edge from } w(u) \text{ to } (0,0,\dots,0,w(u)_{e+d+1}+z)$$
$$\in \hat{H} \text{ is added.}$$

Consequently

$$P_r\{0(d+e+1) \text{ is connected to at least one vertex in } \hat{H} \text{ in } \mathcal{G}(d,e+1)\} = 1. \tag{2.18}$$

By translation, any vertex $(0,0,\dots,z) \in H$ is now connected in $\mathcal{G}(d,e+1)$ to a further vertex $(0,0,\dots,0,z')$ with $z' > z$, a.e. $[P_r]$. From this (2.1) is immediate.

Again we need a minor technical extra step when (2.17) fails. Now we only have

$$\sum_{\substack{v\in\mathbb{Z}^{d+e+1} \\ v_{d+e+1}\geq 0}} r(v) = \infty \tag{2.19}$$

and we can no longer guarantee $z > 0$, respectively $z' > z$ in the last argument. To circumvent this difficulty we think of $\mathcal{G}(d,e+1)$ as being constructed (under P_r) in two stages. Define

$$\hat{r}(v) = 1 - (1-r(v))^{1/2}$$

and choose two independent graphs, $\mathcal{G}'$ and $\mathcal{G}''$ say, each governed by the law $P_{\hat{r}}$. Then the graph $\mathcal{G}$, which includes an edge $\{v,w\}$ if $\{v,w\}$ occurs in at least one of $\mathcal{G}'$ or $\mathcal{G}''$, is governed by the law P_r. We now first construct $\mathcal{H}'$ corresponding to $\mathcal{G}'$ by replacing r by $\hat{r}$ in the preceding proof. One easily checks that $\hat{r}$ also satsifies the analogues of (1.3)–(1.5) as well as $\sum \hat{r}(v) = \infty$. Thus also (2.16) with r replaced by $\hat{r}$ and $\mathcal{G}$ by $\mathcal{G}'$ is valid. Let $v^{(1)},\dots,v^{(k)}$ be a fixed sequence such that

$$r(v^{(i)}) > 0 (\text{hence } \hat{r}(v^{(i)}) > 0), \ \sum_{j\leq i} v^{(j)} \in \mathcal{V}(d,e+1), \ 1 \leq i \leq k,$$

$$v^{(1)} + \dots + v^{(k)} = \xi_{d+e+1}.$$

Then conditionally on $\mathcal{H}'$, for each $u \in \mathcal{V}(d,e)$, there is still a probability of

$$s(u,z) := \{\prod_{j=1}^{k} \hat{r}(v^{(j)})\}\hat{r}((0,0,...,0,w(u)_{e+d+1}+1+z)-(w(u)+\xi_{e+d+1}))$$

that all the following edges are present in $\mathcal{G}''$:

$$\{w(u)+\sum_{j<i} v^{(j)}, w(u)+\sum_{j<i+1} v^{(j)}\},\ 1 \le i \le k,$$

$$\{w(u)+\sum_{j\le k} v^{(j)}, (0,0,...,0,w(u)_{e+d+1}+1+z)\} =$$

$$\{w(u)+\xi_{e+d+1}, (0,0,...,0,w(u)_{e+d+1}+1+z)\}.$$

If all these edges are present in $\mathcal{G}''$, then $w(u)$ is connected in $\mathcal{G}'' \cap \mathcal{V}(d,e+1)$, and hence also in $\mathcal{G}(d,e+1)$, to $(0,0,...,0,w(u)_{e+d+1}+1+z)$. But (by (2.19) and (1.4))

$$\sum_{\substack{z\ge 0\\ u\in\mathcal{V}(d,e)}} s(u,z) = \prod_{1}^{k} \hat{r}(v^{(j)}) \sum_{\substack{z\ge 0\\ u\in\mathcal{V}(d,e)}} \hat{r}(-u_1,...,-u_{e+d},z) = \infty.$$

Thus, w.p.1 in $\mathcal{H}'$ and $\mathcal{G}''$ together, and hence in $\mathcal{G}(d,e+1)$, there will be infinitely many u for which $w(u)$ is connected to $0(d+e+1)$ (in $\mathcal{H}'$) and to

$$\{(0,0,...,w(u)_{e+d+1}+1+z) : z \ge 0\} \subset \hat{H}$$

(in $\mathcal{G}''$). Thus we again have (2.18) and we end the proof of (2.1) as before. □

With the lemma in hand, the **proof of the Theorem** can now be completed essentially by copying [KW], Theorem 2. The proof of (1.6) for $\mathcal{G}(d,e)$ is the same as for $\mathcal{G}(d,0)$, so that we only have to prove (1.7). Assume that (1.7) has already been proven for some $d, e \ge 0$, and let $r : \mathbb{Z}^d \times \mathbb{Z}_+^{e+1}$ satisfy the analogues of (1.3)−(1.5) with e replaced by $(e+1)$. Then by the lemma (2.1) holds. Note that this is even true when $d = e = 0$ by the Borel-Cantelli lemma. Indeed, in this case

$$\sum_{v\in H} r(v) = \sum_{v\in\mathcal{V}(0,1)} r(v) \ge \frac{1}{2}\sum_{v\in\mathbb{Z}} r(v) = \infty.$$

Now for $a, b \in \mathbb{Z}_+$ write $a \overset{n}{\leftrightarrow} b$ if there exists a path in $\mathcal{G}(d,e+1) \cap (\mathbb{Z}^d \times \mathbb{Z}_+^e \times [0,n])$ from $(0,...,0,a)$ to $(0,...,0,b)$ ($d+e$ zeroes here). Then $\overset{n}{\leftrightarrow}$ is an equivalence relation and hence partitions $[0,n]$ into components. Denote the (random) number of these components by X_n. As in [KW] there exists a subsequence $n_1 < n_2 < ...$ such that (E_r denotes expectation with respect to P_r)

$$\liminf_{k\to\infty} E_r\{X_{n_k+a} - X_{n_k}\} \ge 0.$$

Also

$$E_r\{[X_{n+a}-X_n]^+\}$$

$$\leq \sum_{j=1}^{a} P_r\{(0,...,n+j) \text{ is not connected to any } (0,...,b) \text{ with}$$

$$0 \leq b \leq n \text{ in } \mathbb{Z}^d \times \mathbb{Z}_+^e \times [0,n+a]\}$$

$$\leq aP_r\{0(d+e+1) \text{ is not connected to any } (0,...,b) \text{ with}$$

$$a \leq b \leq n \text{ in } \mathbb{Z}^d \times \mathbb{Z}_+^e \times [0,n]\} \to 0 \ (n\to\infty),$$

by virtue of (2.1). Thus

$$E_r\{[X_{n_k+a}-X_{n_k}]^-\} = -E_r\{X_{n_k+a}-X_{n_k}\} + E_r\{[X_{n_k+a}-X_{n_k}]^+\} \to 0 \ (k\to\infty).$$

Now choose a vertex $v \in \mathcal{V}(d,e+1)$ with $r(v)>0$ and $v_{d+e+1} \geq 1$ (such a v exists by (1.5)). If we write $\bar{v} = (-v_1,...,-v_{d+e},v_{d+e+1})$ then $r(\bar{v}) = r(v) > 0$ by (1.3) and (1.4). We take $a = 2v_{d+e+1}$ so that $v+\bar{v} = (0,...,0,a)$. Next we let $\tilde{A}(n)$ be the event that there is some $b \in [0,n)$ such that $(n+a) \overset{(n+a)}{\longleftrightarrow} b$, but there is *no* path from $(0,...,0,n)$ to $(0,...,b)$ in $\mathcal{G}(d,e+1) \cap \{\mathbb{Z}^d \times \mathbb{Z}_+^e \times [0,n]\}$. In addition $(n+a) \overset{(n+a)}{\longleftrightarrow} n$, and for each $n<c<n+a$ there exists a $0 \leq d \leq n$ such that $c \overset{(n+a)}{\longleftrightarrow} d$. On the event $\tilde{A}(n)$ there exists a $b \in [0,n)$ such that $(n+a) \overset{(n+a)}{\longleftrightarrow} b$ and the components of $(0,...,0,n)$ and $(0,...,0,b)$ in $\mathcal{G}(d,e+1) \cap \{\mathbb{Z}^d \times \mathbb{Z}_+^e \times [0,n]\}$ are distinct, while $(0,...,0,n)$ and $(0,...,0,b)$ are connected (via $(0,...,0,n+a)$) in $\mathcal{G}(d,e+1) \cap \{\mathbb{Z}^d \times \mathbb{Z}_+^e \times [0,n+a]\}$. It follows that $X_{n+a} \leq X_n - 1$ on $\tilde{A}(n)$ and

$$P_r\{\tilde{A}(n_k)\} \leq E_r\{[X_{n_k+a}-X_{r_k}]^-\} \to 0$$

as $k\to\infty$. Denote by $A(n)$ the reflection of $\tilde{A}(n)$ in the hyperplane $\{x_1 = (n+a)/2\}$, i.e.,

$$A(n) = \{\text{for some } a < b \leq n+a, 0 \overset{(n+a)}{\longleftrightarrow} b \text{ but there is no path from}$$

$$(0,...,0,a) \text{ to } (0,...,0,b) \text{ in } \mathcal{G}(d,e+1) \cap \{\mathbb{Z}^d \times \mathbb{Z}_+^e \times [a,n+a]\}.$$

$$\text{In addition } 0 \overset{(n+a)}{\longleftrightarrow} a, \text{ and for each } 0<c<a \text{ there exists a}$$

$$d \in [a,n+a] \text{ such that } c \overset{(n+a)}{\longleftrightarrow} d\}.$$

Then by symmetry also

$$P_r\{A(n_k)\} \to 0, k\to\infty.$$

But one easily sees that $A(n)$ occurs on the intersection of the following four events:

$B :=$ the edges $\{0(d+e+1), v\}$ and $\{v, v+\bar{v}\}$ are present in $\mathcal{G}(d, e+1)$,

$C(n) :=$ there is no path from $(0, ..., 0, a)$ to $0(d+e+1)$ in $\mathcal{G}(d, e+1) \cap \{\mathbb{Z}^d \times \mathbb{Z}_+^e \times [0, n+a]\}$ which uses neither of the edges $\{0(d+e+1), v\}$ or $\{v, v+\bar{v}\}$,

$D(n) :=$ for some $b \in (a, n+a]$ there is a path from $0(d+e+1)$ to $(0, ..., 0, b)$ in $\mathcal{G}(d, e+1) \cap \{\mathbb{Z}^d \cap \mathbb{Z}_+^e \cap [0, n+a]\}$ which uses neither of the edges $\{0(d+e+1), v\}, \{v, v+\bar{v}\}$.

$E(n) :=$ for each $0 < c < a$ there exists a $d \in [a, n+a]$ such that $c \overset{(n+a)}{\longleftrightarrow} d$.

We therefore have

$$P_r\{B \cap C(n_k) \cap D(n_k) \cap E(n_k)\} \to 0.$$

Note that P_r {both edges $\{0(d+e+1), v\}$ and $\{v, v+\bar{v}\}$ are *absent* from $\mathcal{G}(d, e+1)\} > 0$. Therefore, by (2.1)

$$P_r\{D(n) \cap E(n)\} \to 1.$$

Moreover, by definition, B and $C(n_k)$ are independent so that

$$P_r\{B \cap C(n_k)\} = r(v)r(\bar{v})P_r\{C(n_k)\} \to 0.$$

Finally (with C^c = complement of C)

$$P_r\{0(d+e+1) \leftrightarrow (0, ..., 0, a) \text{ in } \mathcal{G}(d, e+1)\} \geq \lim_{k \to \infty} P_r\{C^c(n_k)\} = 1. \tag{2.20}$$

From (2.20) we obtain by translation that for any $w \in \mathcal{V}(d, e+1)$

$$P_r\{w \leftrightarrow w + (0, ..., 0, ka) \text{ for all } k \geq 1 \text{ in } \mathcal{G}(d, e+1)\} = 1.$$

In addition a simple Borel-Cantelli argument shows that for any $w', w'' \in \mathcal{V}(d, e+1)$

$$w' + (0, ..., 0, ka) \leftrightarrow w'' + (0, ..., 0, ka) \text{ in } \mathcal{G}(d, e+1) \text{ for infinitely many } k \text{ w.p.1.}$$

Thus

$$P_r\{\mathcal{G}(d, e+1) \text{ is connected}\} = 1.$$

Since we know (1.7) to be true for $e = 0$ (by [GKM], [AKN] or [GKN]), (1.7) follows for all e by induction.

References

[AKN] Aizenman, M., Kesten, H., and Newman, C.M. (1987) Uniqueness of the infinite cluster and continuity of connectivity functions for short and long range percolation, *Comm. Math. Phys.* 111, 505-531.

[BGN] Barsky, D.J., Grimmett, G. and Newman, C.M. (1989), in preparation.

[GKM] Grimmett, G.R., Keane, M. and Marstrand, J.M. (1984) On the connectedness of a random graph, *Math. Proc. Camb. Phil. Soc.* 96, 151-166.

[GKN] Gandolfi, A., Keane, M.S. and Newman, C.M. (1989) Uniqueness of the infinite component in a random graph with applications to percolation and spin glasses, Preprint, Technical University, Delft.

[GM] Grimmett, G. and Marstrand, J.M. (1989), in preparation.

[KW] Kalikow, S. and Weiss, B. (1988) When are random graphs connected?, *Isr. J. Math.* 62, 257-268.

One dimensional Schrödinger operators with stationary deterministic potentials

Shinichi Kotani

1. We are interested in the operator $H(q)$ given by

$$(1.1) \qquad H(q)u = -u'' + qu$$

$$(1.2) \qquad H(q)u_n = u_{n+1} + u_{n-1} + q_n u_n$$

on $L^2(\mathbb{R})$ and $l^2(\mathbb{Z})$, respectively. $\{q(x)\}_{x\in\mathbb{R}}$ and $\{q_n\}_{n\in\mathbb{Z}}$ are assumed to be stationary ergodic processes on a probability space $(\Omega, \mathcal{F}, P)$. In this article we give a brief survey for these operators in the case the potentials q are deterministic. Let

$$\mathcal{F}_- = \sigma-\{q(x); x \le 0\}, \ \mathcal{F}_+ = \sigma-\{q(x); x \ge 0\}.$$

Definition 1.1. $\{q(x)\}_{x\in\mathbb{R}}$ is *deterministic* if $\mathcal{F}_- = \mathcal{F}_+$ (mod.P). It is *non-deterministic* if not.

One of the fundamental results to approach the above problem is the following Herglotz functions method. We say a holomorphic function h on the upper half plane $\mathbb{C}_+$ is of Herglotz if $\operatorname{Im} h > 0$ on $\mathbb{C}_+$. Let

$$g_\lambda^\pm(x, y, q) = (H_\pm(q) - \lambda)^{-1}(x, y)$$

for $\lambda \in \mathbb{C}\backslash\mathbb{R}$, where $H_\pm(q)$ are self-adjoint operators $H(q)|_{\mathbb{R}_\pm}$ with Neumann boundary condition at the boundary point 0. Then it is well-known that $g_\lambda^\pm(0,0,q)$ are of Herglotz.
Set

$$h_\pm(\lambda, q) = -g_\lambda^\pm(0,0,q)^{-1}.$$

Then they are also of Herglotz. We must introduce the Lyapunov exponent. Let $U(x, q, \lambda)$ be the fundamental matrix for

$$U'(x) = \begin{pmatrix} 0 & , & 1 \\ q-\lambda & , & 0 \end{pmatrix} U(x).$$

Define

$$\gamma(\lambda) = \lim_{x\to\infty} \frac{1}{x} \log \| U(x, q, \lambda) \|$$

Probability Theory

if it exists. The subadditive ergodic theorem shows the existence of γ for *a.e.* q and the ergodicity of $\{q(x)\}$ implies its independence of q. $\gamma(\lambda)$ is called the Lyapunov exponent. It is clear that $\gamma(\lambda) \geq 0$. Now it is already proved that

Theorem 1.2. (see [5]) *$\{\lambda \in \mathbb{R}; \gamma(\lambda) = 0\} = \sum_{a.c}(q)$ for a.e. q, where $\sum_{a.c}(q)$ is the absolutely continuous spectrum for $H(q)$.*

Theorem 1.3. ([5]) *For a.e. q,*

$$h_+(\lambda + i0, q) = -\overline{h_-(\lambda + i0, q)}$$

for a.e. $\lambda \in \{\lambda \in \mathbb{R}; \gamma(\lambda) = 0\}$.

It is known that any Herglotz function h has a finite limit $h(\lambda + i0)$ on $\mathbb{R}$ for *a.e.* λ and that h can be recovered completely from the knowledge of $h(\lambda + i0)$ on any given subset $\sum$ of $\mathbb{R}$ with positive Lebesgue measure. Therefore, by combining with a continuity argument, the above two theorems imply

Theorem 1.4. ([5]) *Suppose $|\{\lambda \in \mathbb{R}; \gamma(\lambda) = 0\}| > 0$. Then $\{q(x)\}$ is deterministic. Moreover if $\{q(x), P\}$ are realized on $\Omega = L^2_{real}(\mathbb{R}, \frac{dx}{1+|x|^3})$, there exist continuous $\phi_\pm : \Omega_\pm \longrightarrow \Omega_\mp$ which are also bijective, where*

$$\Omega_\pm = \{q|_{\mathbb{R}_\pm}; q \in \Omega\} \cap \operatorname{supp} P.$$

Theorem 1.2 and 1.4 imply that if $\{q(x)\}$ is non-deterministic, then $H(q)$ has no absolutely continuous spectrum for *a.e.* q. Therefore it is interesting to study the deterministic cases.

2. A little consideration shows

Theorem 2.1. ([4]) *Let P_1, P_2 be two shift invariant probability measures on Ω. Suppose* supp $P_1 \subset$ supp P_2. *Then $\{\gamma_{P_1}(\lambda) = 0\} \supset \{\gamma_{P_2}(\lambda) = 0\}$.*

This especially says that the absolutely continuous spectrum can be determined only by the support of P in $\Omega = L^2_{real}(\mathbb{R}, \frac{dx}{1+|x|^3})$. This theorem has several applications. Let $\{X_x(\omega)\}$ be a stationary ergodic Gaussian process and $f(x)$ be a bounded continuous funciton on $\mathbb{R}$ which is not a constant.
Set

$$q(x) = f(X_x(\omega)).$$

Then $H(q)$ has no absolutely continuous spectrum with probability one. This is easily seen if we notice that in this case supp $P = \{q$: measurable on $\mathbb{R}$ satisfying $\inf f \leq q(x) \leq \sup f$ *a.e.* $x \in \mathbb{R}.\}$. Another example is given in [3] and the potentials are of the form:

$$q(x) = \sum_{n \in \mathbb{Z}} f(x+n)q_n$$

If f has not a compact support, then $q(x)$ may be deterministic even though $\{q_n\}$ is independent and identically distributed. However, under suitable conditions on f , one can show the absence of the absolutely continuous spectrum. These two cases are sufficiently random although they are deterministic, and almost periodic potentials are not included.

3. Suppose

$$q(x) = \sum_{n \in \mathbb{Z}} I_{[n,n+1)}(x) q_n$$

where $\{q_n\}$ is a stationary ergodic sequence taking only finitely many values. Then we have

Theorem 3.1. ([6]) *Unless $\{q_n\}$ is periodic, $H(q)$ has no absolutely continuous spectrum with probability one.*

So far we have been considering only the case (1.1). However we have analogous theorems also for (1.2) (see [10]). The corresponding theorem is of particular interest when

$$q_n = Q(n\alpha + \omega) \ ,$$

where $\alpha \in \mathbb{R}^d$, $\omega \in \mathbb{R}^d/\mathbb{Z}^d$ and Q is a finitely many valued function on $\mathbb{R}^d/\mathbb{Z}^d$. Historically the following Fibonacci case has been intensively studied; namely

$$\alpha = \frac{\sqrt{5}-1}{2}, \ Q(\omega) = \kappa I_{[1-\alpha,1]}(\omega) \ .$$

In this respect, recently there appeared two papers which prove

Theorem 3.2. ([8],[9]) *Suppose α is irrational and $\kappa \neq 0$. Then $\sum = \{\lambda \in \mathbb{R}; \gamma(\lambda) = 0\}$ is closed and coincides with the spectrum of $H(\omega)$. For any $\omega \in \mathbb{R}^d/\mathbb{Z}^d$, $H(\omega)$ has no absolutely continuous spectrum and in particular $H(0)$ has only singular continuous spectrum. Moreover $|\sum| = 0$, and hence $\sum$ has empty interior.*

We remark that $|\sum| = 0$ is an easy consequence of Theorem 3.1. [8] has proved only in the case $\alpha = \frac{\sqrt{5}-1}{2}$.

If the almost periodic potentials take infinitely many values, then the situation becomes completely different. In (1.1), we assume

$$q(x) = \kappa\{\cos 2\pi x + \cos 2\pi(\alpha x + \omega)\}$$

where α is irrational and $\omega \in \mathbb{R}/\mathbb{Z}$. Then it is known that

Theorem 3.3. ([1],[2],[7]) *Suppose κ is sufficiently large and α satisfies*

$$|n|^2 |\sin 2\pi \alpha n| \geq c > 0 \ for \ any \ n \in \mathbb{Z} \ .$$

Then for a.e. $\omega \in \mathbb{R}/\mathbb{Z}$, $H(\omega)$ has purely point spectrum in a certain neighbourhood of the bottom of the spectrum and there exists also absolutely continuous spectrum.

To prove this theorem one must use the KAM theory quite delicately.

As we see in the above series of theorems, it is quite interesting to investigate more deeply almost periodic Schrödinger operators.

References

[1] Dinaburg, E., Sinai, Ya, One dimensional Schrödinger equation with quasi-periodic potential, *Functional Anal. Appl.* **9** (1975), 279.

[2] Fröhlich, J., Spencer, T. & Wittwer, P., Localization for a class of one-dimensional quasi-periodic potentials, *Comm. Math. Phys.* **132** (1990), 5-25.

[3] Kirsch, W., Kotani, S. & Simon, B., Absence of absolutely continuous spectrum for some one-dimensional random but deterministic Schrödinger operators, *Ann. Inst. H. Poincaré* **42** (1985), 383.

[4] Kotani, S., Support theorems for random Schrödinger operators, *Comm. Math. Phys.* **97** (1985), 443.

[5] Kotani, S., One dimensional Schrödinger operators and Herglotz functions, *Proc. Taniguchi Symp. Katata and Kyoto 1985*, ed. Itô, K., Ikeda, N.

[6] Kotani, S., Jacobi matrices with random potentials taking finitely many values, *Review in Math. Phys.*, Vol. 1 (1990), 129-133.

[7] Sinai, Ya., Anderson localization for one-dimensional difference Schrödinger operator with quasi-periodic potential, *J. Statist. Phys.* **46** (1987), 861.

[8] Sütö, A., Singular continuous spectrum on a Cantor set of zero Lebesgue measure for the Fibonacci Hamiltonian, to appear in *J. Statist. Phys.*

[9] Bellissard, J., Jochum, B., Scoppola, E. & Testard, D., Spectral properties of one-dimensional quasi-crystals, to appear in *Comm. Math. Phys.*

[10] Simon, B., Kotani theory for one-dimensional stochastic Jacobi matrices, *Comm. Math. Phys.* **89** (1983), 227.

Lyapunov exponents and rotation numbers of linear systems with real noise*

Mark A. Pinsky and Volker Wihstutz

1. Introduction

A large literature has been devoted to studying the asymptotic properties of the linear stochastic system

$$\dot{X}_t = AX_t + \varepsilon BX_t F(\xi_t), \quad X_0 = x \in R^d \tag{1.1}$$

where A, B are constant $d \times d$ matrices and $F(\xi_t)$ is a mean-zero function of an ergodic Markov process on a compact state space M. Of particular interest is the top Lyapunov exponent

$$\lambda(\varepsilon) = \lim_{t \uparrow \infty} t^{-1} \log|X_t| \tag{1.2}$$

and the rotation number, suitably defined in case $d = 2$. To analyze this system it is noted that the joint process (X_t, ξ_t) is Markovian on the product space $R^d \times M$ with infinitesimal generator

$$L = L_X = G + AX \cdot \nabla + \varepsilon(BX \cdot \nabla)F(\xi) \tag{1.3}$$

Here G is the generator of the noise process (ξ_t); the second term is a "systematic derivative," making no reference to the noise process. The third term mixes the noise and state variables, inviting the term "noisy derivative" and is the source of the analytical challenge. This model is referred to as a "real-noise driven system."

It is well known that the noise process obeys a central limit theorem in the form

$$\delta \int_0^{t/\delta^2} F(\xi_s)ds \Rightarrow N(0, \sigma^2 t) \quad (\delta \downarrow 0) \tag{1.4}$$

a normal law with mean zero and variance $\sigma^2 t$, where the variance parameter $\sigma^2 = -2\langle G^{-1}F, F\rangle$ and the inner product is computed in terms of the invariant measure of the process (ξ_t). Therefore one may attempt to analyze the asymptotics of (X_t, ξ_t) by studying a related system driven by *white noise*

$$dY_t = AY_t dt + \varepsilon BY_t \circ dw_t \tag{1.5}$$

---

*Research supported by Air Force Office of Scientific Research

Probability Theory

where (w_t) is a Wiener process with mean zero and variance $\sigma^2 t$. The diffusion process (Y_t) is Markovian on R^d and has infinitesimal generator

$$L_Y = AY \cdot \nabla + \frac{1}{2}\varepsilon^2\sigma^2(BY \cdot \nabla)^2. \tag{1.6}$$

One may conjecture that the asymptotic behaviour of the Lyapunov exponent for the real-noise driven process (X_t) is equivalent to that for the diffusion process (Y_t), at least to the first approximation. It is our purpose to carry out the details of this plan in cases of interest.

2. Nilpotent systems

In a previous paper [5] we investigated the Lyapunov exponent for white noise systems with a nilpotent deterministic part: $A^d = 0, A^{d-1}$ non-zero and B generic. This includes the free particle perturbed by multiplicative white noise as well as other models of physical interest. Recently these results were extended to systems driven by "telegraphic noise," where $M = \{-1, 1\}$, by Arnold and Kloeden [7]. Now we can show that these results can be extended to the general real-noise driven system. We have the following theorem.

Theorem 2.1. *Let $\lambda(\varepsilon), \tau(\varepsilon)$ be the Lyapunov exponent and rotation number of the 2×2 system $\dot{X}_t = [A + \varepsilon BF(\xi_t)]X_t$ where $A^2 = 0, A$ non-zero and (ξ_t) is a finite-state ergodic Markov process. Suppose that $\langle Be^{\perp}, e\rangle > 0$ where $Ae = 0, e$* non-zero, $\langle e, e^{\perp}\rangle = 0$. *When $\varepsilon \downarrow 0$ we have*

$$\lambda(\varepsilon) \sim C_1\varepsilon^{2/3}, \quad \tau(\varepsilon) \sim C_2\varepsilon^{2/3} \tag{2.1}$$

for positive constants C_1, C_2. These expansions are precisely the same as for the Lyapunov exponent and rotation number of the associated diffusion process (Y_t).

We compute the Lyapunov exponent by the "adjoint method." This consists in writing the generator in terms of a system of polar coordinates (ρ, φ) and setting $Q(\xi, \varphi) = L(\rho)$. The angular process (ξ_t, φ_t) is ergodic with stationary measure $N(d\xi \times d\varphi)$ and the usual formula for the Lyapunov exponent is $\lambda(\varepsilon) = \int_{M \times S^{d-1}} Q(\xi, \varphi)N(d\xi \times d\varphi)$. In the present case, $\lambda(\varepsilon)$ may be characterized as the unique number λ for which there exists a function $f(\varphi, \xi)$ solving the equation $Lf = Q - \lambda$. Indeed, integrating both sides against $N(d\xi \times d\varphi)$ shows that $\lambda = \lambda(\varepsilon)$. Since this equation may be difficult to solve exactly, we may obtain asymptotic approximations by replacing L by a suitable approximate generator, or equivalently to find a function f_ε and a number λ_ε such that $Lf_\varepsilon = Q - \lambda_\varepsilon + O(R_\varepsilon)$ for a suitable remainder term R_ε. Integrating this equation against $N(d\xi \times d\varphi)$ produces the asymptotic statement $\lambda_\varepsilon = \lambda(\varepsilon) + O(R_\varepsilon)$. It remains to find the approximations $f_\varepsilon, \lambda_\varepsilon, R_\varepsilon$.

To do this we apply a method of "homogenization." We write the noisy part of the generator in the form $G + \delta V$ and show that this is approximated by a diffusion operator in the following sense: there exists a second order operator $L_o : C^\infty(R^2) \to C^\infty(R^2)$ and operators $L_i : C^\infty(R^2) \to C^\infty(R^2 \times M)(i = 1,2)$ such that for each $f \in C^\infty(R^2)$ we have

$$(G + \delta V)(f + \delta L_1 f + \delta^2 L_2 f) = \delta^2 (L_o f) + O(\delta^3) \quad (\delta \downarrow 0). \tag{2.2}$$

This allows us to reduce to the case of a white-noise driven system for which we know the asymptotics of the Lyapunov exponent and rotation number. The details appear in sections 4 and 5.

3. The harmonic oscillator

We now compare the small-noise behaviour of the stochastic harmonic oscillator for the cases of real noise and white noise. The real noise system is defined by $\dot{X}_t = AX_t dt + \varepsilon BX_t F(\xi_t)$ where $A = \begin{pmatrix} 0 & 1 \\ -\gamma & 0 \end{pmatrix}$ and $B = \begin{pmatrix} 0 & 0 \\ 1 & 0 \end{pmatrix}$ with $\gamma > 0$. We assume specifically that (ξ_t) is a finite-state Markov process with self-adjoint generator (reversible case) and invariant measure $\nu(d\xi)$. We take a system of polar coordinates (ρ, φ) with $x_1\sqrt{\gamma} = e^\rho \cos\varphi, x_2 = e^\rho \sin\varphi$. The stochastic equations take the form

$$\begin{aligned} \dot{\varphi}_t &= -\sqrt{\gamma} + \varepsilon[F(\xi_t)/\sqrt{\gamma}]\cos^2\varphi_t \\ \dot{\rho}_t &= \varepsilon[F(\xi_t)/\sqrt{\gamma}]\sin\varphi_t \cos\varphi_t . \end{aligned} \tag{3.1}$$

The infinitesimal generator of the joint motion of $(\xi_t, \varphi_t, \rho_t)$ is

$$L_X = G - \sqrt{\gamma}\frac{\partial}{\partial\varphi} + \varepsilon[F(\xi)/\sqrt{\gamma}](\cos^2\varphi\frac{\partial}{\partial\varphi} + \sin\varphi\cos\varphi\frac{\partial}{\partial\rho}).$$

The white noise system is defined by the Stratonovich equation

$$dY_t = AY_t dt + \varepsilon BY_t \circ dw_t \tag{3.2}$$

where w_t is a Wiener process with mean zero and variance $\sigma^2 t$. Its generator is given by

$$L_Y = -\sqrt{\gamma}\frac{\partial}{\partial\varphi} + \frac{\varepsilon^2\sigma^2}{2\gamma}(\cos^2\varphi\frac{\partial}{\partial\varphi} + \sin\varphi\cos\varphi\frac{\partial}{\partial\rho})^2$$

where we make the identification $\sigma^2 = -2\langle G^{-1}F, F\rangle$.

To obtain the asymptotic form of the Lyapunov exponent of the real-noise driven system we look for an approximate solution (f, λ) of the equation $L_X f = Q - \lambda$ where $Q(\xi, \varphi) = L\rho = \varepsilon[F(\xi)/\sqrt{\gamma}]\sin\varphi \cos\varphi$. This is sought in the form

$$\begin{aligned} f &= f_0 + \varepsilon f_1 + \varepsilon^2 f_2 \\ \lambda &= \lambda_0 + \varepsilon\lambda_1 + \varepsilon^2\lambda_2 \end{aligned} \tag{3.3}$$

leading to the conditions

$$(G - \sqrt{\gamma}\frac{\partial}{\partial\varphi}) f_0 = \lambda_0$$

$$(G - \sqrt{\gamma}\frac{\partial}{\partial\varphi}) f_1 + [F(\xi)/\sqrt{\gamma}]\cos^2\varphi f_0' = [F(\xi)/\sqrt{\gamma}]\sin\varphi\cos\varphi - \lambda_1$$

$$(G - \sqrt{\gamma}\frac{\partial}{\partial\varphi}) f_2 + [F(\xi)/\sqrt{\gamma}]\cos^2\varphi f_1' = -\lambda_2,$$

choosing $\lambda_0 = 0$ this leads to $f_0 = 0, \lambda_1 = 0, f_1 = (G - \sqrt{\gamma}\frac{\partial}{\partial\varphi})^{-1} F(\xi)\sin\varphi\cos\varphi/\sqrt{\gamma}$ which may be solved in terms of eigenfunctions ψ_k where $G\psi_k = -\mu_k\psi_k$ for $k = 1, \dots, N$ with $\psi_1 = 1, \mu_1 = 0$ and $\mu_k > 0$ for $k > 1$ by writing $F(\xi)\sin\varphi\cos\varphi = \frac{1}{2}\sum_{k>1}\langle F, \psi_k\rangle\psi_k\sin 2\varphi$, leading to

$$f_1 = (1/\sqrt{\gamma})\sum_{k>1}[\frac{1}{2}\mu_k\sin 2\varphi + \sqrt{\gamma}\cos 2\varphi]\langle F, \psi_k\rangle\psi_k/(\mu_k^2 + 4\gamma).$$

Averaging the f_2 equation with respect to the normalized measure $d\varphi\nu(d\xi)$ and integrating by parts leads to

$$\begin{aligned} -\lambda_2 &= (1/\sqrt{\gamma})\int F(\xi)\cos^2\varphi f_1' d\varphi\nu(d\xi) \\ &= -(2/\sqrt{\gamma})\int F(\xi)\sin\varphi\cos\varphi f_1 d\varphi\nu(d\xi). \end{aligned}$$

From the above spectral representations, we can compute the inner product of f_1 and $F\sin 2\varphi$ to obtain

$$\int f_1 F(\xi)\sin\varphi\cos\varphi d\varphi\nu(d\xi) = \sum_{k>1}\mu_k\langle F, \psi_k\rangle^2\langle\sin^2 2\varphi\rangle/4({\mu^2}_k + 4\gamma)$$

with the result ([4],[6])

$$\lambda_2^{\mathrm{real}} = (1/4\gamma)\sum_{k>1}\mu_k\langle F, \psi_k\rangle^2/(\mu_k^2 + 4\gamma). \tag{3.4}$$

To obtain the asymptotic form of the Lyapunov exponent of the white-noise driven system we look for an approximate solution (f, λ) of the equation $L_Y f = Q - \lambda$ where $Q(\varphi) = L_Y\rho = (\varepsilon^2\sigma^2/2\gamma)\cos^2\varphi\cos 2\varphi$. This is sought in the form

$$f = f_0 + \varepsilon^2 f_2$$

$$\lambda = \lambda_0 + \varepsilon^2\lambda_2$$

leading to the conditions

$$-\sqrt{\gamma} f_0' = \lambda_0$$

$$\sqrt{\gamma} f_2' + \frac{\sigma^2}{2\gamma}(\cos^2\varphi \frac{\partial}{\partial\varphi})^2 f_0 = (\sigma^2/2\gamma)\cos^2\varphi\cos 2\varphi - \lambda_2$$

leading to the choices $\lambda_0 = 0, f_0 = 0$ and the well known result (replacing $d\varphi$ by $d\varphi/2\pi$)

(3.5)
$$\lambda_2^{\text{white}} = (\sigma^2/2\gamma)(2\pi)^{-1} \int_{-\pi}^{\pi} \cos^2\varphi\cos 2\varphi d\varphi = (\sigma^2/8\gamma).$$

If we make the identification $\sigma^2 = -2\langle G^{-1}F, F\rangle$, then this may be written as

(3.6)
$$\lambda_2^{\text{white}} = (1/4\gamma) \sum_{k>1} \langle F, \psi_k\rangle^2/\mu_k.$$

These computations are summarized as

Proposition 3.1. *We always have the inequality* $\lambda_2^{\text{real}} < \lambda_2^{\text{white}}$.

To resolve this apparent discrepancy, it suffices to consider a parametrized family of real-noise processes $\int_0^t \delta^{-1}F(\xi_{s\delta^{-2}})ds$. When $\delta \downarrow 0$ these converge to a Wiener process with mean zero and variance $\sigma^2 = -2\langle G^{-1}F, F\rangle$. The stochastic equation has the form

$$\dot{X}_t = AX_t + \epsilon[\delta^{-1}F(\xi_{t\delta^{-2}})]BX_t$$

with infinitesimal generator

$$L_X = AX\cdot\nabla + \delta^{-2}G + \epsilon\delta^{-1}F(\xi)BX\cdot\nabla.$$

To obtain the corresponding form of the Lyapunov exponent, it suffices to substitute above, with μ_k replaced by $\mu_k\delta^{-2}$ and F replaced by $F\delta^{-1}$. Thus

$$\lambda_2^{\text{real}}(\delta) = (1/4\gamma)\delta^{-2} \sum_{k>1} (\mu_k\delta^{-2})\langle F, \psi_k\rangle^2/[(\mu_k\delta^{-2})^2 + 4\gamma] = (1/4\gamma) \sum_{k>1} \mu_k\langle F, \psi_k\rangle^2/[\mu_k^2 + 4\gamma\delta^4].$$

When $\delta \downarrow 0$ we have

Proposition 3.2. $\lim_{\delta\downarrow 0} \lambda_2^{\text{real}}(\delta) = (1/4\gamma) \sum_{k>1} \langle F, \psi_k\rangle^2/\mu_k = \lambda_2^{\text{white}}$.

Thus we retrieve the white noise result in the CLT limit.

4. Proof of theorem 2.1 (special case)

We are given an ergodic Markov process $\{\xi(t)\}_{t\geq 0}$ on a compact state space M; the infinitesimal generator is denoted G and the invariant measure ν – thus $G^*\nu = 0$ and $G1 = 0$. We further assume that the Fredholm alternative is satisfied for the simple eigenvalue zero, i.e., the inhomogeneous equation $Gf = g$ has a solution provided that $\int_M g(\xi)\nu(d\xi) = 0$; the solution is uniquely determined by requiring $\int_M f(\xi)\nu(d\xi) = 0$. This condition is satisfied for a finite-state Markov process or for Brownian motion on a compact manifold, for example.

Let there be given a function $F(\xi)$ with mean value zero, i.e., $\int_M F(\xi)\nu(d\xi) = 0$. Let $(x(t), \dot{x}(t)) = (x_1(t), x_2(t))$ be the solution of the second-order system $\ddot{x}(t) = \varepsilon x(t)F(\xi(t))$ with the initial conditions $x(0) = x_1, \dot{x}(0) = x_2$. This is a Markov process on the product space $R^2 \times M$ with the infinitesimal generator

$$L = G + x_2\frac{\partial}{\partial x_1} + \varepsilon F(\xi)x_1\frac{\partial}{\partial x_2}. \tag{4.1}$$

The top Lyapunov exponent is defined by

$$\lambda(\varepsilon) = \lim_{t\downarrow\infty} t^{-1}\log\sqrt{x_1(t)^2 + x_2(t)^2}. \tag{4.2}$$

This is invariant under linear change of coordinates in (x_1, x_2) space, in particular the scaling transformation $(x_1, x_2) \to (Cx_1, x_2)$.

We introduce a system of "polar coordinates" by

$$x_1 = e^{\rho}\cos\varphi, \quad x_2 = Ce^{\rho}\sin\varphi. \tag{4.3}$$

We make the identification $x(t) = x_1(t), \dot{x}(t) = x_2(t)$ and consider the joint process $(\xi(t), \rho(t), \varphi(t))_{t\geq 0}$. After a short calculation we find that

$$\begin{aligned} \dot{\varphi}(t) &= -C\sin^2\varphi(t) + (\varepsilon/C)\cos^2\varphi(t)F(\xi(t)) \\ \dot{\rho}(t) &= C\sin\varphi(t)\cos\varphi(t) + (\varepsilon/C)\sin\varphi(t)\cos\varphi(t)F(\xi(t)). \end{aligned} \tag{4.4}$$

The joint process $(\xi(t), \varphi(t), \rho(t))_{t\geq 0}$ is a Markov process on the space $M \times R \times R$ with the infinitesimal generator

$$\begin{aligned} L &= G + \varepsilon F(\xi)x_1\frac{\partial}{\partial x_2} + x_2\frac{\partial}{\partial x_1} \\ &= G + (-C\sin^2\varphi + (\varepsilon/C)\cos^2\varphi F(\xi))\frac{\partial}{\partial\varphi} \\ &\quad + \sin\varphi\cos\varphi(C + (\varepsilon/C)F(\xi))\frac{\partial}{\partial\rho} \end{aligned} \tag{4.5}$$

(in fact the first two components already form a Markov process, but we shall need the full generator in what follows).

We write the generator in the form $L = G + \delta V + D$, where

$$V = F(\xi)(\sin\varphi\cos\varphi\frac{\partial}{\partial\rho} + \cos^2\varphi\frac{\partial}{\partial\varphi})$$

$$D = C\sin\varphi\cos\varphi\frac{\partial}{\partial\rho} - C\sin^2\varphi\frac{\partial}{\partial\varphi}$$

and $\delta = \varepsilon/C$. We refer to G as the *noise generator*, V as the *noisy derivatives* and D as the *systematic derivatives*. We also note, for further reference, the function $Q(\varphi,\xi)$ defined as $L\rho(\varphi,\xi)$ is computed as

$$Q(\varphi,\xi) = C\sin\varphi\cos\varphi + (\varepsilon/C)F(\xi)\sin\varphi\cos\varphi. \tag{4.6}$$

In previous approaches to stochastic Lyapunov stability, one solves approximately the equation $Lf = Q - \lambda$ for suitable $f = f(\varphi,\xi)$ and $\lambda \in R$. In the present case this is not directly possible, because of the presence of the noise variables $F(\xi)$ which intervene both in the generator and in $Q(\varphi,\xi)$. Therefore we apply a process of "homogenization" to replace the operator $G+\delta V$ by a suitable diffusion operator in the (φ,ρ) space and ultimately replace $Q(\varphi,\xi)$ by a function $Q(\varphi)$ related to a suitable diffusion process.

Proposition 4.1. *There exists a second-order differential operator L_0 in the (φ,ρ) variables with the following property: for any $f \in C^\infty(R\times R)$ there exist correctors $f_i \in C^\infty(R\times R\times M)(i = 1,2)$ such that*

$$\begin{aligned}&[G+\delta V](f+\delta f_1+\delta^2 f_2)\\ &=: [G + (\delta F(\xi)(\cos^2\varphi\frac{\partial}{\partial\varphi} + \sin\varphi\cos\varphi\frac{\partial}{\partial\rho})](f+\delta f_1+\delta^2 f_2)\\ &= \delta^2(L_o f)(\varphi,\rho) + O(\delta^3) \quad (\delta\downarrow 0)\end{aligned} \tag{4.7}$$

where $L_o f(\rho,\varphi) = \frac{1}{2}\sigma^2(\sin\varphi\cos\varphi\frac{\partial}{\partial\rho} + \cos^2\varphi\frac{\partial}{\partial\varphi})^2 f$ and $\sigma^2 = -2\langle G^{-1}F, F\rangle > 0$. The term $O(\delta^3)$ is estimated in terms of the C^3 sup norm of f.

Proof. We choose the correctors in order to cancel the term of order δ and to render the coefficient of δ^2 independent of ξ. This requires that we have the equations

$$Gf_1 + \cos^2\varphi F(\xi)\frac{\partial f}{\partial\varphi} + \sin\varphi\cos\varphi F(\xi)\frac{\partial f}{\partial\rho} = 0 \tag{4.8}$$

$$Gf_2 + \cos^2\varphi F(\xi)\frac{\partial f_1}{\partial\varphi} + \sin\varphi\cos\varphi F(\xi)\frac{\partial f_1}{\partial\rho} = \text{function of } (\varphi,\rho). \tag{4.9}$$

The first of these is satisfied by taking

$$f_1 = H(\xi)[\cos^2\varphi\frac{\partial f}{\partial\varphi} + \sin\varphi\cos\varphi\frac{\partial f}{\partial\rho}]$$

where $H(\xi)$ is the solution of $GH = -F$, normalized so that $\int_M H(\xi)\nu(d\xi) = 0$. With this choice of f_1 we substitute in the equation (4.9) for f_2 and average with respect to $\nu(d\xi)$. The Gf_2 term drops out and the right side of the equation is found to be

$$(4.10) \qquad \frac{1}{2}\sigma^2(\cos^2\varphi\frac{\partial}{\partial\varphi} + \sin\varphi\cos\varphi\frac{\partial}{\partial\rho})^2 f =: L_o f$$

where $\sigma^2 = 2\int_M F(\xi)H(\xi)\nu(d\xi)$.

Finally f_2 is determined by solving the indicated equation (4.9) subject to the normalization $\int_M f_2(\xi,\varphi,\rho)\nu(d\xi) = 0$. This is possible, since Gf_2 has been arranged to be perpendicular to the null space of G^*, completing the proof. □

If $\varepsilon/C \downarrow 0$ we may restrict attention to the "approximate generator" $(\varepsilon/C)^2 L_o f - C\sin^2\varphi\frac{\partial}{\partial\varphi} + C\sin\varphi\cos\varphi\frac{\partial}{\partial\rho}$. In order for the terms to balance we are led to the equation $(\varepsilon/C)^2 = C$ or $C = \varepsilon^{2/3}$ as we had in the diffusion case [5]. More precisely, we consider the white noise system $dx_1 = x_2 dt, dx_2 = \varepsilon x_1 \circ dw$ where $\{w(t) : t \geq 0\}$ is a Wiener process with mean zero and variance parameter $\sigma^2 t$. For this system the infinitesimal generator is $\tilde{L}_\varepsilon = x_2\frac{\partial}{\partial x_1} + \varepsilon^2\frac{\sigma^2}{2}(x_1\frac{\partial}{\partial x_2})^2$; if we take the polar coordinate system $x_1 = Ce^\rho\cos\varphi, x_2 = e^\rho\sin\varphi$, we obtain the angular equation

$$(4.11) \qquad d\varphi = -C\sin\varphi\cos\varphi + (\sigma\varepsilon/C)^2\cos^2\varphi \circ dw$$

with the Q function ($Q_\varepsilon = \tilde{L}_\varepsilon\rho$)

$$(4.12) \qquad Q_\varepsilon(\varphi) = C\sin\varphi\cos\varphi + \frac{1}{2}(\sigma\varepsilon/C)^2\cos^2\varphi\cos 2\varphi.$$

From our previous work [5], we know that if we choose $C = \varepsilon^{2/3}$, then $\tilde{L}_\varepsilon = \varepsilon^{2/3}\tilde{L}_1$, where $\tilde{L}_1$ is hypoelliptic with invariant measure μ and the Lyapunov exponent is $\lambda_1 = \int_{-\pi}^{\pi} Q_1(\varphi)\mu(d\varphi) > 0$.

In order to find the Lyapunov exponent of the real-noise driven system, it suffices to find $f(\varphi,\xi)$ and $\lambda^\sim$ such that $Lf(\varphi,\xi) = Q(\varphi,\xi) - \lambda^\sim + O(\varepsilon)$. To do this, we proceed in three steps.

Step 1. Let $\rho_1(\varphi,\xi), \rho_2(\varphi,\xi)$ be the correctors such that

$$(4.13) \qquad (G + \delta V)(\rho + \delta\rho_1 + \delta^2\rho_2) = \delta^2 L_o\rho + O(\delta^3).$$

In our previous paper [5] we showed that the operator $-\sin^2\varphi\frac{\partial}{\partial\varphi} + L_0$ on $[-\pi,\pi]$ has an invariant measure μ and satisfies the Fredholm alternative for the simple eigenvalue zero and that $\lambda_1 =: \int_{-\pi}^{\pi} Q(\varphi)\mu(d\varphi) > 0$, where $Q(\varphi) =: \sin\varphi\cos\varphi + L_o\rho = \sin\varphi\cos\varphi + \frac{1}{2}\sigma^2\cos^2\varphi\cos 2\varphi$.

Step 2. Let $h = h(\varphi)$ be the solution of the equation

$$(4.14) \qquad (-\sin^2\varphi\frac{\partial}{\partial\varphi} + L_o)h = Q(\varphi) - \lambda_1$$

normalized so that $\int_{-\pi}^{\pi} h(\varphi)\mu(d\varphi) = 0$.

Step 3. Let $h_1(\varphi,\xi), h_2(\varphi,\xi)$ be the correctors defined above for the function h, i.e., $(G+\delta V)(h+\delta h_1+\delta^2 h_2) = \delta^2 L_o h + O(\delta^3)$.

Proposition 4.2. *With the above notations, we let*

$$f(\varphi,\xi) = h(\varphi) + \delta(h_1-\rho_1)(\varphi,\xi) + \delta^2(h_2-\rho_2)(\varphi,\xi) \tag{4.15}$$

where $\delta = \varepsilon/C, C = \varepsilon^{2/3}$. *Then* $Lf(\varphi,\xi) = Q(\varphi,\xi) - \varepsilon^{2/3}\lambda_1 + O(\varepsilon)$ *in particular the Lyapunov exponent* $\lambda(\varepsilon) = \varepsilon^{2/3}\lambda_1 + O(\varepsilon), \varepsilon \downarrow 0$.

Proof. Recall that $L = G + \delta V + D$ where G is the noise generator, V contains the noisy derivatives and D contains the systematic derivatives. From step 1, we have

$$\begin{aligned}(G+\delta V)(\rho+\delta\rho_1+\delta^2\rho_2) &= \frac{1}{2}\delta^2\sigma^2\cos^2\varphi\cos 2\varphi + O(\varepsilon)\\ L(\rho+\delta\rho_1+\delta^2\rho_2) &= C\sin\varphi\cos\varphi + \frac{1}{2}\delta^2\sigma^2\cos^2\varphi\cos 2\varphi + O(\varepsilon)\\ &= \varepsilon^{2/3}Q(\varphi) + O(\varepsilon)\end{aligned}$$

$$\begin{aligned}(G+\delta V)(h+\delta h_1+\delta^2 h_2) &= \delta^2 L_o h + O(\varepsilon)\\ L(h+\delta h_1+\delta^2 h_2) &= -C\sin^2\varphi h'(\varphi) + \delta^2 L_o h + O(\varepsilon)\\ &= \varepsilon^{2/3}(Q(\varphi)-\lambda_1) + O(\varepsilon).\end{aligned}$$

Subtracting these two gives

$$L(h-\rho+\delta(h_1-\rho_1)+\delta^2(h_2-\rho_2)) = -\varepsilon^{2/3}\lambda_1 + O(\varepsilon).$$

Adding $Q(\varphi,\xi) = L\rho$ to both sides gives the desired result. □

To complete the proof, we may argue directly in terms of martingales: writing the above equation as $L\tilde{f} = -\varepsilon^{2/3}\lambda_1 + O(\varepsilon)$, we have that $M_t =: \tilde{f}(\rho(t),\varphi(t),\xi(t)) - \int_0^t L\tilde{f}(\rho(s),\varphi(s),\xi(s))ds$ is a martingale. But $M_t = o(t)$ when $t\uparrow\infty$; therefore we may divide by t and take the limit obtaining $\lim_{t\uparrow\infty}\rho(t)/t = -\lambda_1\varepsilon^{2/3} + O(\varepsilon)$. We have the following theorem.

Theorem 2.1(A). *The top Lyapunov exponent satisfies*

$$\lambda(\varepsilon) = \lim_{t\uparrow\infty} t^{-1}\rho(t) = \varepsilon^{2/3}\langle\sin\varphi\cos\varphi + \frac{1}{2}\sigma^2\cos^2\varphi\cos 2\varphi\rangle_\mu + O(\varepsilon) \quad (\varepsilon\downarrow 0).$$

By the same method we can compute the rotation number, defined as $\tau(\varepsilon) = \lim_{t\uparrow\infty} t^{-1}\varphi(t)$. To do this we verify the following:

Proposition 4.3. *Let $k(\varphi)$ be the solution of*

$$-\sin^2\varphi k'(\varphi) + L_o k = L_o\varphi - \sin^2\varphi - \langle L_o\varphi\rangle_\mu + \langle\sin^2\varphi\rangle_\mu.$$

Let $k_1(\varphi), k_2(\varphi)$ be the correctors defined above for $f = \varphi + k(\varphi)$. Then $L(\varphi + k(\varphi) + \varepsilon^{1/3}k_1(\varphi) + \varepsilon^{2/3}k_2(\varphi)) = \varepsilon^{2/3}\langle L_o\varphi + \sin^2\varphi\rangle_\mu + O(\varepsilon)$.

Noting that $k(\varphi)$ and the correctors $\varphi_1, \varphi_2, k_1, k_2$ are periodic functions we have the following result:

Theorem 2.1(B). *The rotation number is computed as*

$$\tau(\varepsilon) = \lim_{t\uparrow\infty}\varphi(t)/t = \varepsilon^{2/3}\langle\sin^2\varphi - \frac{1}{2}\sigma^2\cos^2\varphi\sin 2\varphi\rangle_\mu + O(\varepsilon) \quad (\varepsilon\downarrow 0).$$

5. Proof of theorem 2.1 (general case)

We now generalize the set-up of the previous section to the stochastic system

$$\dot{x}(t) = [A + \varepsilon F(\xi(t))B]x(t) \tag{5.1}$$

where $x(t) = (x_1(t), x_2(t))$ and $A \neq 0$ is a 2×2 matrix with $A^2 = 0$. Without loss of generality we may take a basis in which $A = \begin{pmatrix} 0 & 1 \\ 0 & 0 \end{pmatrix}$ so that the system has the form

$$\begin{aligned} \dot{x}_1(t) &= x_2(t) + \varepsilon F(\xi(t))(b_{11}x_1(t) + b_{12}x_2(t)) \\ \dot{x}_2(t) &= \varepsilon F(\xi(t))(b_{21}x_1(t) + b_{22}x_2(t)). \end{aligned} \tag{5.2}$$

Taking a system of polar coordinates $x_1 = Ce^\rho\cos\varphi, x_2 = e^\rho\sin\varphi$ we find that

$$\begin{aligned} \dot{\varphi} &= -C\sin^2\varphi - \varepsilon\sin\varphi(b_{11}\cos\varphi + b_{12}C\sin\varphi)F(\xi) \\ &\quad + (\varepsilon/C)\cos\varphi(b_{21}\cos\varphi + b_{22}C\sin\varphi)F(\xi) \\ \dot{\rho} &= C\sin\varphi\cos\varphi + \varepsilon\cos\varphi(b_{11}\cos\varphi + b_{12}C\sin\varphi)F(\xi) \\ &\quad + (\varepsilon/C)\sin\varphi(b_{21}\cos\varphi + b_{22}C\sin\varphi)F(\xi). \end{aligned}$$

Making the choice $C = \varepsilon^{2/3}$, the generator has the form

$$\begin{aligned} L &= \varepsilon^{2/3}\{-\sin^2\varphi\frac{\partial}{\partial\varphi} + \sin\varphi\cos\varphi\frac{\partial}{\partial\rho}\} \\ &\quad + G + \varepsilon^{1/3}V_1 + \varepsilon V_2 + \varepsilon^{5/3}V_3 \end{aligned}$$

where

$$V_1 = b_{21}\cos^2\varphi F(\xi)\frac{\partial}{\partial\varphi} + b_{21}F(\xi)\sin\varphi\cos\varphi\frac{\partial}{\partial\rho}$$

$$V_2 = (b_{22} - b_{11})\sin\varphi\cos\varphi F(\xi)\frac{\partial}{\partial\varphi} + (b_{11}\cos^2\varphi + b_{22}\sin^2\varphi)F(\xi)\frac{\partial}{\partial\rho}$$

$$V_3 = -b_{12}\sin^2\varphi\frac{\partial}{\partial\varphi} + b_{12}\sin\varphi\cos\varphi\frac{\partial}{\partial\rho}.$$

This has the same structure as in the case of white noise. If the coefficient b_{21} is non-zero, we may apply the homogenization procedure described above to obtain the approximate generator as

$$\varepsilon^{2/3}b_{21}^2 L_o f + \varepsilon^{2/3}\{-\sin^2\varphi\frac{\partial}{\partial\varphi} + \sin\varphi\cos\varphi\frac{\partial}{\partial\rho}\} + O(\varepsilon).$$

Applying the adjoint method we again obtain the result in the form as stated in Theorem 2.1.

References

[1] *Lyapunov Exponents*, ed. by L. Arnold & V. Wihstutz. Springer-Verlag Lecture Notes in Mathematics, vol. 1186, 1985.

[2] L. Arnold, G. Papanicolaou & V. Wihstutz, Asymptotic analysis of the Lyapunov exponent and rotation number of the random oscillator and applications, *SIAM J. Appl. Math.* **46** (1986), 427-450.

[3] E. Pardoux & V. Wihstutz, Lyapunov exponent and rotation number of two-dimensional stochastic systems with small diffusion, *SIAM J. Appl. Math.* **48** (1988), 442-457.

[4] M. Pinsky, Instability of the harmonic oscillator with small noise, *SIAM J. Appl. Math.* **46** (1986), 451-463.

[5] M. Pinsky & V. Wihstutz, Lyapunov exponents of nilpotent Itô systems, *Stochastics* **25** (1988), 43-57.

[6] V. Wihstutz, Analytic expansion of the Lyapunov exponent associated to the Schroedinger operator with random potential, *Stochastic Anal. Appl.* **3** (1985), 93-118.

[7] L. Arnold & P. Kloeden, Lyapunov exponents and rotation numbers of two-dimensional systems with telegraphic noise, *SIAM J. Appl. Math.*, to appear.

Some results about test and generalized functionals of white noise*

Jürgen Potthoff and Jia-an Yan

1. Introduction

In the recent years the construction of spaces of smooth and generalized functions in infinite dimensions as well as their applications and the investigation of their properties has received increasing interest. Typically the spaces considered are spaces formulated in terms of functionals of Brownian motion [KU 89, Me 83, MY87, Su 85, Us 87, Wa 83] or of white noise [Hi 75, HPS 88, KT 80, KY 87, Ku 83, MY 88, Po 87, Po 88] or of their respective generalizations. Among the motivations of the above quoted articles – besides the purely theoretical interest in these structures – are applications to the Malliavin calculus [Su 85, Wa 83], to the study of partial stochastic differential equations [Ch 88, Ku 83, NZ 88], to the construction and analysis of Dirichlet forms in infinite dimensions [HPS 88, AHP 89], to the generalization of stochastic analysis [Kb 83, KS 87], to a mathematical treatment of the Feynman integral [HS 83, Ku 83a] etc.

The present article is concerned with a special choice of the class of Gel'fand triples

$$(E)^* \supset (L^2) \supset (E) \tag{1.1}$$

which were constructed in [KT 80]: we shall consider the triple

$$(S)^* \supset (L^2) \supset (S) \tag{1.2}$$

where $(L^2) = L^2(S'(\mathbf{R}), \mathcal{B}, d\mu)$, $\mathcal{B}$ being the σ-algebra on $S'(\mathbf{R})$ generated by the weak topology and $d\mu$ is the white noise measure on $(S'(\mathbf{R}), \mathcal{B})$ (e.g. [Hi 80]). The space (S) in (1.2) is the projective limit (e.g. [GV 64])

$$(S) := \operatorname*{proj-lim}_{p \in \mathbf{N}_0} (S)_p \tag{1.3}$$

of Hilbert space $(S)_p$, $p \geq 0$, which are by definition the (completed) (L^2)-domains of the operator $\Gamma(A)^p$. Here $\Gamma(A)$ is the second quantization (e.g. [Ne 73]) of the self-adjoint operator A, which is the (unique) extension of $-\frac{1}{2}\Delta + u^2 + 1$ on $L^2(\mathbf{R}, du)$. (Note that $\Gamma(A)$ is self-adjoint, so that $\Gamma(A)^p$ makes sense.) Moreover,

* This work was supported in part by the A.v.Humboldt-Stiftung, Deutsche Forschungsgemeinschaft and the National Natural Science Foundation of China.

Probability Theory

in (1.2) we have denoted the dual of (S) by $(S)^*$ (and – as usual – identified (L^2) with its dual).

The triple (1.2) has been investigated in a number of papers (e.g. [AHP 89, KY 87, MY 89, HPS 88, Yo 87]). By standard theory (e.g. [GV 64, Tr 67]) (S) is a Fréchet space and

$$(1.4) \qquad (S)^* = \cup_{p\in \mathbf{N}_0}(S)_{-p}$$

where $(S)_{-p} := (S)_p^*$. Moreover, an estimate found in [KT 80] together with the fact that $spec(A) = \{2(k+1); k \in \mathbf{N}_0\}$ show that (S) is nuclear.

In this article we shall prove a number of new results about the spaces (S) and $(S)^*$. Moreover we provide some new proofs of known results (e.g. the property that (S) is an algebra [KT 80]).

In section 2 we shall show first that pointwise multiplication is a continuous operation on (S). Detailed proofs of this fact can be found already in [Yo 87, HPS 88a]. Here we prove this by deriving an estimate for the relevant norms of products of elements in (S) which is sharper than those found in the quoted articles. Next we prove that (S) is embedded continously into the space of test functionals considered by Mayer [Me 83] and Watanabe [Wa 83]. (A similar result has been found independently in [Ku 89].) Then we consider translations of elements in (S) in *arbitrary* directions of $S'(\mathbf{R})$ and show that they act continously on (S). This entails a number of interesting results concerning the translation of measures on $(S'(\mathbf{R}), \mathcal{B})$ which are represented by a positive element in $(S)^*$ in the sense of [Yo 87]. For example, the white noise measure μ_y translated by $y \in S'(\mathbf{R})$ is represented by a positive functional in $(S)^*$ and one may talk of a *generalized Radon-Nikodym derivative* of μ_y with respect to μ in $(S)^*$. Moreover we discuss the scaling transformation of elements in (S) and – again – some consequences for measures on $(S'(\mathbf{R}), \mathcal{B})$ which are given by elements in $(S)^*$.

Section 3 is devoted to the study of differential operators acting on (S). It is proven that the Gâteaux derivative D_y in direction of $y \in S'(\mathbf{R})$ acts continously on (S) for every $y \in S'(\mathbf{R})$. In particular, identifying the Hida derivative ∂_t, $t \in \mathbf{R}$, with D_{δ_t}, we find that (S) is C^∞ under the Hida derivative. Also, we prove that $(S)_\alpha$ belongs to the (L^2) - domain of ∂_t, $t \in \mathbf{R}$, whenever $\alpha > \frac{1}{4}$. Finally we show that Taylor's formula holds in (S) and in particular also pointwise for the pointwise defined (strongly continous) representatives of elements in (S).

The paper is concluded by section 4 which contains a result about the convolution of measures on $(S'(\mathbf{R}), \mathcal{B})$: if two measures are represented by elements in $(S)^*$, then so is their convolution.

Acknowledgements. We acknowledge gratefully the warm hospitality extended to us by Professors S. Albeverio, Ph. Blanchard, L. Streit at BiBoS, University of Bielefeld and by Professor Hu Guo-ding, Nankai Institute of Mathematics, Tianjin. Also, it is our pleasure to thank Professors P.A. Meyer, M. Röckner and L. Streit for stimulating and fruitful discussions. The second author would like to thank Professor Louis H.Y. Chen for his kind invitation and warm hospitality at the Singapore probability Conference, where this paper was presented. He also acknowledges the financial support by the Organization Committee of the conference and the Natural Science Foundation of China.

2. Properties of (S)

This section is devoted to the study of certain properties of the space of test functionals (S).

Note that (L^2) is isomorphic (as a Hilbert space) to the symmetric Fock space over $L^2(\mathbf{R})$ (e.g. [Hi 80, Ne 73, Si 75]), i.e. every element $\varphi \in (L^2)$ is in one-to-one correspondence with a sequence $(f^{(n)}, n \in \mathbf{N}_0), f^{(n)} \in \widehat{L^2(\mathbf{R}^n)}$ ($\hat{\bullet}$ stands for symmetrization), such that

$$\|\varphi\|_2^2 = \sum_{n=0}^{\infty} n! \, \|f^n\|^2_{L^2(\mathbf{R}^n)}. \tag{2.1}$$

We call the function $f^{(n)}$ the *kernel functions of* φ.

The isomorphism between (L^2) and the symmetric Fock space can be implemented by the following integral transformation S (cf. [KT 80])

$$S\varphi(\xi) = \int \varphi(x+\xi)\, d\mu(x), \xi \in S(\mathbf{R}). \tag{2.2}$$

In fact, we have the formula

$$S\varphi(\xi) = \int \varphi(x) : e^{\langle x,\xi\rangle} : d\mu(x) \tag{2.3}$$

where we used the notation $: e^{\langle x,\xi\rangle} :\equiv e^{\langle x,\xi\rangle - \frac{1}{2}\|\xi\|_2^2}$, and $\|\cdot\|_2$ denotes the norm of $L^2(\mathbf{R})$. Upon expansion of the exponential on the right hand side of (2.3), one obtaines an expansion of $S_\varphi(\xi)$ into homogeneous functionals of ξ of degree n, which can be identified with $f^{(n)}$ – viewed as an element in $(L^2(\mathbf{R}))^{\hat{\otimes} n}$ – evaluated at $\xi^{\hat{\otimes} n}$.

Since $: e^{\langle \cdot,\xi\rangle} :$ belongs to (S) for every $\xi \in S(\mathbf{R})$, we can extend the definition of the S-transform consistently to elements $\Phi \in (S)^*$ via

$$S\Phi(\xi) = \langle \Phi, : e^{\langle \cdot,\xi\rangle} : \rangle$$

where the exterior brackets denote the dual pairing of $(S)^*$ with (S).

$\varphi \in (L^2)$ can be represented directly by the kernel functions $f^{(n)}$ as follows:

$$\varphi(x) = \sum_{n=0}^{\infty} \langle : x^{\otimes n} :, f^{(n)} \rangle \tag{2.4}$$

where $:x^{\otimes n}: \in \widehat{S'(\mathbf{R}^n)}$ is the *Wick-ordering* of the distribution $x^{\otimes n} \in \widehat{S'(\mathbf{R}^n)}$, whose definiton we now recall.

Let $x \in S'(\mathbf{R})$, so that $x^{\otimes n} \in \widehat{S'(\mathbf{R}^n)}$, $n \in \mathbf{N}$, by the nuclear theorem. Since sums of the form $\sum_k a_k \xi_{k_1} \hat{\otimes} \cdots \hat{\otimes} \xi_{k_n}$, $k = (k_1, ..., k_n) \in \mathbf{N}^n$, $\xi_{k_j} \in S(\mathbf{R})$, with only a finite number of non-zero terms are dense in $\widehat{S(\mathbf{R}^n)}$, it is sufficient to define $\langle :x^{\otimes n}:, \xi_1 \hat{\otimes} \cdots \hat{\otimes} \xi_n \rangle$. By polarization we then only need to give $\langle :x^{\otimes n}:, \xi^{\hat{\otimes} n} \rangle$, and this is defined by

$$\langle :x^{\otimes n}:, \xi^{\hat{\otimes} n} \rangle :=: \langle x, \xi \rangle^n :_{\sigma^2}, \quad \sigma^2 = \|\xi\|_2^2 \tag{2.5}$$

where the right hand side of (2.5) denotes the Hermite polynomial of order n and parameter σ^2. We have

$$\langle :x^{\otimes n}:, \xi^{\hat{\otimes} n} \rangle = \frac{d^n}{dt^n} \exp\left(t\langle x, \xi \rangle - \frac{1}{2} t^2 \|\xi\|_2^2 \right) \Big|_{t=0}. \tag{2.6}$$

Working out the relation (2.6), one finds the following recursion

$$:x^{\otimes 0}: = 1 \tag{2.7a}$$

$$:x^{\otimes 1}: = x \tag{2.7b}$$

$$:x^{\otimes n}: =: x^{\otimes n-1} : \otimes x - (n-1) : x^{\otimes n-2} : \otimes Tr \tag{2.7c}$$

where we denote by Tr the following distribution in $\widehat{S'(\mathbf{R}^2)}$:

$$\langle Tr, f \rangle = \int_{\mathbf{R}} dt \langle \delta_t^{\otimes 2}, f \rangle = \int_{\mathbf{R}} dt f(t,t) \tag{2.8}$$

$f \in \widehat{S(\mathbf{R}^2)}$. For example:

$$\langle :x^{\otimes 2}:, f \rangle = \langle x^{\otimes 2}, f \rangle - \int_{\mathbf{R}} dt f(t,t).$$

As it stands, (2.4) is slightly informal, because in general if $\varphi \in (L^2)$, we do not have $f^{(n)} \in \widehat{S(\mathbf{R}^n)}$, but $f^{(n)} \in \widehat{L^2(\mathbf{R}^n)}$ and therefore the pairing has in general no meaning. However, if $\varphi \in (S)$ then for every $n \in N$, $f^{(n)}$ has a version in $\widehat{S(\mathbf{R}^n)}$, so that for this version the dual pairing $\langle \cdot, \cdot \rangle$ under the sum (2.4) is well-defined and one can show [KY 87] that the sum over n converges pointwise for every $x \in S'(\mathbf{R})$, uniformly on strongly bounded sets of $S'(\mathbf{R})$. (In fact, Kubo and Yokoi show this only on bounded sets of the inductive limit topology of $S'(\mathbf{R})$ [KY 87], but an application of the Mackey-Ahrens theorem proves that the strong and the inductive limit topology on $S'(\mathbf{R})$ coincide [AHP 89c].) Therefore $\varphi \in (S)$ has a version $\tilde{\varphi}$ which is pointwise defined and strongly continuous, so that for $\varphi \in (S)$ (2.4) may be interpreted pointwise. (Strong continuity of $\tilde{\varphi}$ implies that this version is unique.)

Since (S) is dense in (L^2), for general $\varphi \in (L^2)$ (2.4) has to be interpreted as being the visualization of the (L^2)-limit of functionals in (S).

Note that (2.4) is nothing but the white noise version of the decomposition of φ into multiple Wiener integrals of degree n. Moreover, if φ is represented as in (2.4) we have

$$S\varphi(x) = \sum_{n=0}^{\infty} \int_{R^n} f^{(n)}(u)\xi^{\hat{\otimes} n}(u)du. \tag{2.9}$$

Now we shall prove the following theorem, which has been announced in [KT 80].

Theorem 2.1. (S) *is an algebra (under pointwise multiplication).*

Detailed proofs can be found in [HPS 88a] and [Yo 87]. Here we present a simple proof which seems to give the sharpest result. Theorem 2.1 is a consequence of

Proposition 2.2. *Let* $\varphi, \psi \in (S)_p, p > q_0 \equiv \frac{\ln(2+\sqrt{2})}{2\ln 2}$. *Then for every* $q > q_0$, $\varphi \cdot \psi \in (S)_{p-q}$ *and there exists a constant* $C_q > 0$ *(independent of p) such that*

$$\|\varphi \cdot \psi\|_{2,p-q} \le C_q \|\varphi\|_{2,p} \|\psi\|_{2,p} \tag{2.10}$$

where $\|\cdot\|_{2,p}$ *denotes the norm of* $(S)_p$.

Proof. Let $\varphi(\psi)$ be given by $(f^{(n)}, n \in \mathbf{N}_0)$ $((g^{(n)}, n \in \mathbf{N}_0)$ resp.). By the multiplication formula of Wick-ordered products (e.g. [AP 88, HPS 88a, KT 80]) we have for $m, n \in \mathbf{N}_0$

$$\varphi^{(m)}(x)\psi^{(n)}(x) = \sum_{k=0}^{m\wedge n} k! \binom{m}{k}\binom{n}{k} \langle : x^{\otimes(n+m-2k)} :, f^{(m)}\hat{\otimes}_k g^{(n)} \rangle \tag{2.11}$$

where $\varphi^{(m)}$ ($\psi^{(n)}$ resp.) stand for the n-th (m-th resp.) homogeneous component $\langle : x^{\otimes m} : f^{(m)} \rangle$ ($\langle : x^{\otimes n} :, g^{(n)} \rangle$ resp.) of φ (ψ resp.) and $f^{(m)}\hat{\otimes}_k g^{(n)}$ is the symmetrization of

$$\int_{\mathbf{R}^k} f^{(m)}(u_1, ..., u_k, \cdot) g^{(n)}(u_1, ..., u_k, \cdot) d^k u. \tag{2.12}$$

Since the decomposition (2.4) "diagonalizes" the operator $\Gamma(A)$, we have for $r \in \mathbf{R}$

$$(\Gamma(A)^r \varphi^{(m)}\psi^{(n)})(x) = \sum_{k=0}^{m\wedge n} k! \binom{m}{k}\binom{n}{k} \times$$

(2.13)

$$\times \langle : x^{\otimes(n+m-2k)} :, (A^{\otimes(n+m-2k)})^r f^{(m)} \hat{\otimes}_k g^{(n)} \rangle$$

$$\equiv \sum_{k=0}^{m\wedge n} k! \binom{m}{k} \binom{n}{k} \chi^{(n+m-2k)}(x)$$

(where $A^{\otimes n}$ acts naturally on $h^{(n)} \in L^2(\mathbf{R}^n)$ due to the isomorphism $L^2(\mathbf{R}^n) \simeq L^2(\mathbf{R})^{\otimes n}$). Thus for $r \geq 0$ we may estimate as follows.

$$\begin{aligned}
\|\varphi^{(m)}\psi^{(n)}\|_{2,r} &= \|\Gamma(A)^r \varphi^{(m)}\psi^{(n)}\|_2 \\
&\leq \sum_{k=0}^{n\wedge m} k! \binom{m}{k}\binom{n}{k} \|\chi^{(n+m-2k)}\|_2 \\
&= \sum_{k=0}^{n\wedge m} k! \binom{m}{k}\binom{n}{k} \|(A^{\otimes(n+m-2k)})^r f^{(m)} \hat{\otimes}_k g^{(n)}\|_2 \\
&\leq \sum_{k=0}^{n\wedge m} k! \binom{m}{k}\binom{n}{k} ((n+m-2k)!)^{1/2} \|(A^{\otimes m})^r f^{(m)}\|_2 \times \\
&\qquad \times \|(A^{\otimes n})^r g^{(n)}\|_2
\end{aligned}$$

using Schwarz' inequality and the spectral theorem in the last step (recall that $A \geq 2$). Thus

$$\|\varphi^{(m)}\psi^{(n)}\|_{2,r} \leq \|\varphi^{(m)}\|_{2,r} \|\psi^{(n)}\|_{2,r} \cdot \sum_{k=0}^{n\wedge m} [\binom{m}{k}\binom{n}{k}\binom{n+m-2k}{n-k}]^{1/2}.$$

Now we use $\binom{m}{k}\binom{n}{k} \leq \binom{m+n}{2k}$ and $\binom{n+m-2k}{n-k} \leq 2^{n+m-2k}$ and therefore obtain

$$\begin{aligned}
\sum_{k=0}^{n\wedge m} [\binom{m}{k}\binom{n}{k}\binom{n+m-2k}{n-k}]^{\frac{1}{2}} &\leq 2^{\frac{(m+n)}{2}} \sum_{k=0}^{m+n} \binom{m+n}{2k}^{\frac{1}{2}} 2^{\frac{-k}{2}} \\
&\leq 2^{\frac{(m+n)}{2}} (\sum_{k=0}^{\infty} 2^{\frac{-k}{2}})^{-\frac{1}{2}} (\sum_{k=0}^{m+n} \binom{m+n}{2k} 2^{-\frac{k}{2}})^{\frac{1}{2}} \\
&= (\frac{\sqrt{2}}{\sqrt{2}-1})^{\frac{1}{2}} (\sqrt{2+\sqrt{2}})^{m+n}.
\end{aligned}$$

Altogether we have the estimate

$$\|\varphi^{(m)}\psi^{(n)}\|_{2,r} \leq \text{const.}\sigma^m \|\varphi^{(m)}\|_{2,r} \cdot \sigma^n \|\psi^{(n)}\|_{2,r}$$

with $\sigma = \sqrt{2+\sqrt{2}} < 2$. Note that $(A^{\otimes n})^{-1} \leq 2^{-n}$ for every n. Therefore

$$\begin{aligned} \|\varphi\psi\|_{2,r} &\leq \sum_{m,n=0}^{\infty} \|\varphi^{(m)}\psi^{(n)}\|_{2,r} \\ &\leq \text{const.}(\sum_{n=0}^{\infty} \sigma^n \|\varphi^{(n)}\|_{2,r})(\sum_{n=0}^{\infty} \sigma^n \|\psi^{(n)}\|_{2,r}) \\ &\leq \text{const.} \sum_{n=0}^{\infty} (\frac{\sigma}{2^q})^n (\sum_{n=0}^{\infty} \|\varphi^{(n)}\|_{2,r+q}^2)^{\frac{1}{2}} \times \\ &\quad \times (\sum_{n=0}^{\infty} \|\psi^{(n)}\|_{2,r+q}^2)^{\frac{1}{2}} \\ &= \text{const.} \|\varphi\|_{2,r+q} \|\psi\|_{2,r+q} \end{aligned}$$

whenever $q > 0$, such that $\frac{\sqrt{2+\sqrt{2}}}{2^q} < 1$, i.e. $q > q_0$. □

Another interesting property of (S) is that it is continuously embedded into the space D of test functions considered by Meyer [Me 83], Watanabe [Wa 83], Sugita [Su 85] and others. Let us quickly recall the definition of D within the white noise framework (cf. also [Po 87,88]).

N denotes the number (or Ornstein-Uhlenbeck) operator which is given by multiplication by $n \in \mathbf{N}_0$ on the n-th homogeneous chaos and is defined as the unique self-adjoint extension of this operation on the algebra P of polynomial functions. For $k \in \mathbf{N}_0$, $p \geq 1$, $\varphi \in P$ set

$$|\varphi|_{p,k} := \|(1+N)^k \varphi\|_p \tag{2.14}$$

where $\|\cdot\|_p$ is the norm of $L^p(S'(\mathbf{R}), B, d\mu)$. Obviously (2.14) defines a norm on P and we put

$$D_{p,k} := \overline{P}^{|\cdot|_{p,k}} \tag{2.15}$$

and define D to be the projective limit of the system $\{D_{k,p}, p \in \mathbf{N}, k \in \mathbf{N}_0\}$. Clearly D is a Fréchet space. Its dual is denoted by D^*.

Theorem 2.3. *$(S) \subset D$, the injection from (S) into D being continuous.*

Proof. It suffices to show that for every $p \geq 1$, $k \geq 0$, $\varphi \in P$ we have the inequality

$$|\varphi|_{p,k} \leq C_{p,q,k} \|\varphi\|_{2,q} \tag{2.16}$$

for some $q > 0$ and a positive constant $C_{p,q,k}$.

Note that $\|A^{-r}\| \le 2^{-r}$ (operator norm on $L^2(\mathbf{R})$), as can be seen from the spectrum of A (s.a.). Therefore Nelson's hypercontractivity theorem [Ne 73] implies that $\Gamma(A^{-r})$ is a contraction from (L^p) into (L^2), whenever $r \ge r_0 \equiv \frac{\ln(p+1)}{\ln 2}$. Then for such $r > r_0$

$$\begin{aligned}|\varphi|_{p,k} &= \|\Gamma(A^{-r_0})\Gamma(A^{r_0})(1+N)^k\varphi\|_p\\ &\le \|\Gamma(A^{r_0})(1+N)^k\varphi\|_{2,p}\\ &= \|(1+N)^k\varphi\|_{2,p+r_0}\\ &\le C_{p,q,k}\|\varphi\|_{2,q}\end{aligned}$$

with $q = p + r$. In the last step we used the inequality $\Gamma(A)^{-s} \le 2^{-sN}$ for every $s \ge 0$, which is a simple consequence of $A \ge 2$ and the definition of $\Gamma(A)$. □

Corollary 2.4.

$$\mathcal{D}^* \subset (\mathcal{S})^*.$$

In particular, composites of elements in $\mathcal{S}'(\mathbf{R}^n)$ with $\mathbf{R}^n$-valued functionals in $\mathcal{D}$ (satisfying the usual nondegeneracy condition, cf. [Wa 83]) can be found as elements in $(\mathcal{S})^*$ according to Watanabe's theorem [Wa 83] (cf. also [Po 88, Su 85]). Moreover, by Yokoi's theorem we know that for positive distributions in $\mathcal{S}'(\mathbf{R}^n)$ these composites are represented (on $(\mathcal{S})$) by a measure on $(\mathcal{S}'(\mathbf{R}), \mathcal{B})$. We shall apply these ideas to the construction of Dirichlet forms over $\mathcal{S}'(\mathbf{R}^d)$ in a forthcoming publication.

Next we turn our attention to the action of translations on $(\mathcal{S})$. Since $\varphi \in (\mathcal{S})$ has an everywhere defined version $\tilde{\varphi}$, we may define for $y \in \mathcal{S}'(\mathbf{R})$ the $\mathcal{S}'(\mathbf{R})$-functional $(\varphi_y)^\sim$ by setting

$$(\varphi_y)^\sim := \tilde{\varphi}(x+y). \tag{2.17}$$

The question arises whether $(\varphi_y)^\sim$ defines an element in (φ), i.e. if there exists $\varphi_y \in (\mathcal{S})$ so that its continuous version coincides with (2.17).

Theorem 2.5. *For every $y \in \mathcal{S}'(\mathbf{R})$ the mapping $\tau_y : \varphi \to \tau_y\varphi = \varphi_y$, defined on $\mathcal{P}$, extends to a continuous mapping from $(\mathcal{S})$ into itself. Moreover $\tilde{\varphi}_y = (\varphi_y)^\sim$.*

Proof. Let $\varphi \in \mathcal{P}$, $\varphi(x) = \sum_n \langle : x^{\otimes n} :, f^{(n)}\rangle$ (sum with finitely many non-zero terms), and identify φ with its continuous version (i.e. identify $f^{(n)} \in \mathcal{S}(\widehat{\mathbf{R}^n})$ with its class in $L^2(\widehat{\mathbf{R}^n})$).

From the relations (2.6) or (2.7) for the Wick-ordered distribution $: x^{\otimes n} :$, $x \in \mathcal{S}'(\mathbf{R})$, it follows by a straightforward computation that for $y \in \mathcal{S}'(\mathbf{R})$

$$: (x+y)^{\otimes n} := \sum_{k=0}^{n} \binom{n}{k} : x^{\otimes k} : \hat{\otimes} y^{\otimes(n-k)}. \tag{2.18}$$

For $k \leq n$ put $f_y^{(n,k)} := \langle y^{\otimes k}, f^{(n)} \rangle \in S(\widehat{\mathbf{R}^{n-k}})$, i.e. $y^{\otimes k}$, acts in the first k variables of $f^{(n)}$. Therefore

$$\Gamma(A)^p \varphi_y(x) = \sum_{n=0}^{\infty} \sum_{k=0}^{n} \binom{n}{k} \langle : x^{\otimes k} :, (A^{\otimes k})^p f_y^{(n,n-k)} \rangle$$

for every p. Let $p \in \mathbf{N}_0$, then

$$\begin{aligned} \|\varphi_y\|_{2,p}^2 &= \sum_{m,n=0}^{\infty} \sum_{k=0}^{n} k! \binom{n}{k} \binom{m}{k} \times \\ &\quad \times \left((A^{\otimes k})^p f_y^{(n,n-k)}, (A^{\otimes k})^p f_y^{(m,m-k)} \right)_{L^2(\mathbf{R}^k)} \\ &\leq \sum_{m,n=0}^{\infty} \sum_{k=0}^{n} k! \binom{n}{k} \binom{m}{k} \|(A^{\otimes k})^p f_y^{(n,n-k)}\|_2 \|(A^{\otimes k})^p f_y^{(m,m-k)}\|_2 . \end{aligned}$$

Schwarz' inequality and the spectral theorem imply that

$$\|(A^{\otimes k})^p f_y^{(n,n-k)}\|_2 \leq \|A^{-(p+q)} y\|_2^{n-k} \|(A^{\otimes n})^{p+q} f^{(n)}\|_2$$

whenever q is chosen large enough so that $A^{-(p+q)} y \in L^2(\mathbf{R})$. Furthermore we use again $\binom{n}{k}\binom{m}{k} \leq \binom{m+n}{2k}$ and the estimate $k! \leq (n!m!)^{1/2}$ under the above sum. Therefore we obtain $(r = p + q)$

$$\begin{aligned} \|\varphi y\|_{2,p}^2 &\leq \sum_{m,n=0}^{\infty} (n!m!)^{1/2} \|(A^{\otimes n})^r f^{(n)}\|_2 \|(A^{\otimes m})^r f^{(m)}\|_2 \times \\ &\quad \times \sum_{k=0}^{m \wedge n} \binom{n+m}{2k} \|A^{-r} y\|_2^{n+m-2k} \\ &\leq \Big(\sum_{n=0}^{\infty} (n!)^{1/2} (1 + \|A^{-r} y\|_2)^n \|(A^{\otimes n})^r f^{(n)}\|_2 \Big)^2 . \end{aligned}$$

Since $(A^{\otimes n})^{-s} \leq 2^{-sn}$ for all $s \geq 0$, an application of Schwarz' inequality gives for s large enough the bound

$$\begin{aligned} \|\varphi_y\|_{2,p}^2 &\leq \text{const.} \sum_{n=0}^{\infty} n! \, \|(A^{\otimes n})^{p+s} f^{(n)}\|_2^2 \\ &= \text{const.} \|\varphi\|_{2,p+s}^2 . \end{aligned}$$

Density of $\mathcal{P}$ in (S) completes the proof of the existence of the continuous extension of τ_y. The last part of the theorem is proven by the remark that the evaluation map $\delta_x : \varphi \to \tilde{\varphi}(x)$ is continuous on (S) for every $x \in S'(\mathbf{R})$ (cf. [KY 87]). □

Remark. At the end of the next section we shall derive the Taylor expansion of $\tau_y\varphi$.

Theorem 2.5 has some interesting consequences. Since $1 \in (S)^*$ (i.e. the expectation belongs to $(S)^*$) we have for every $y \in S'(\mathbf{R})$

$$\begin{aligned}|\langle 1, \varphi_y\rangle| &= |\int \varphi(x) d\mu(x-y)| \\ &\leq \text{const.} \|\varphi\|_{2,q}\end{aligned}$$

for some $q \in \mathbf{N}_0$. On the other hand, if $y \in L^2(\mathbf{R})$

$$\begin{aligned}\langle 1, \varphi_y\rangle &= \int \varphi(x) e^{\langle x,y\rangle - \frac{1}{2}\|y\|_2^2} d\mu(x) \\ &= (: e^{\langle\cdot,y\rangle} :, \varphi)_{(L^2)}.\end{aligned}$$

Therefore we see that, although for general $y \in S'(\mathbf{R})$ $d\mu(\cdot - y)$ is no longer absolutely continuous with respect to $d\mu$, one can realize the translation of the white noise measure by $y \in S'(\mathbf{R})$ as a generalized functional in $(S)^*$. We shall talk of a *generalized Radon-Nikodym derivative* $\frac{d\mu(\cdot-y)}{d\mu}$ of $d\mu(\cdot - y)$ *with respect to* $d\mu$ in $(S)^*$. It is given by : $\exp(\langle\cdot, y\rangle)$:, i.e. it is the generalized functional whose sequence of kernels is given by $((n!)^{-1}y^{\otimes n}, n \in \mathbf{N}_0)$.

More generally we can prove the following result.

Define the *Wick Product* $\Phi : \Psi$ of two elements in $(S)^*$ by $\Phi : \Psi = S^{-1}[S\Phi \cdot S\Psi]$. It is shown in [MY 88] that $\Phi : \Psi \in (S)^*$. In fact, it turns out that if Φ, resp. Ψ, have kernels $(T^{(n)}, n \in \mathbf{N}_0)$, $(U^{(n)}, n \in \mathbf{N}_0)$ resp., then the kernels of $\Phi : \Psi$ are given by the sequence $(\sum_{m=0}^n T^{(n-m)} \hat{\otimes} U^{(m)}, n \in \mathbf{N}_0)$.

Theorem 2.6. *Let ν be a positive finite measure on $(S'(\mathbf{R}), \mathcal{B})$ such that $\frac{d\nu}{d\mu} \in (S)^*$. For $y \in S'(\mathbf{R})$ denote $\nu(\cdot - y)$ by ν_y. Then $\frac{d\nu_y}{d\mu} \in (S)^*$ and one has*

$$\frac{d\nu_y}{d\mu} = \frac{d\mu_y}{d\mu} : \frac{d\nu}{d\mu} \tag{2.19}$$

$$(S\frac{d\mu_y}{d\mu})(\xi) = e^{\langle y,\xi\rangle}. \tag{2.20}$$

Proof. Let $\frac{d\nu}{d\mu} \in (S)_{-p}$, $\varphi \in (S)$. Then

$$\begin{aligned}\int \tilde{\varphi}(x) d\nu_y(x) &= \int \tilde{\varphi}(x+y) d\nu(x) \\ &= \langle \frac{d\nu}{d\mu}, \varphi_y\rangle\end{aligned}$$

and by Theorem 2.5 we find the estimate

$$|\int \tilde{\varphi}(x)d\nu_y(x)| = |\langle \frac{d\nu}{d\mu}, \varphi_y\rangle| \\ \leq \|\frac{d\nu}{d\mu}\|_{2,-p}\|\varphi\|_{2,q}$$

for some $q > p$. Therefore $\frac{d\nu_y}{d\mu} \in (S)_{-p}$ and we have

$$\langle \frac{d\nu_y}{d\mu}, \varphi\rangle = \langle \frac{d\nu}{d\mu}, \varphi_y\rangle. \tag{2.21}$$

The particular choice $\varphi(x) =: \exp(\langle x, \xi\rangle) :, \xi \in S(\mathbf{R})$, gives

$$(S\frac{d\nu_y}{d\mu})(\xi) = \langle \frac{d\nu_y}{d\mu}, : e^{\langle \cdot, \xi\rangle} :\rangle \\ = e^{\langle y, \xi\rangle}\langle \frac{d\nu}{d\mu}, : e^{\langle \cdot, \xi\rangle} :\rangle \\ = e^{\langle y, \xi\rangle}(S\frac{d\nu}{d\mu})(\xi)$$

from which (2.19) and (2.20) follow. □

Corollary 2.7.

$$\frac{d\mu_y}{d\mu} : \frac{d\mu_{-y}}{d\mu} = 1 \tag{2.22}$$

$$\frac{d\nu_y}{d\mu} : \frac{d\nu_{-y}}{d\mu} = \frac{d\nu}{d\mu} : \frac{d\nu}{d\mu}. \tag{2.23}$$

Consider $\Phi \in (S)$, $\eta \in S(\mathbf{R})$, $\varphi \in (S)$. Then

$$\langle \Phi_\eta, \varphi\rangle = \int \phi(x+\eta)\varphi(x)d\mu(x) \\ = \int \Phi(x)\varphi(x-y) : e^{\langle x, \eta\rangle} : d\mu(x) \\ = (\Phi, : e^{\langle \cdot, \eta\rangle} : \varphi_{-\eta}).$$

Upon taking limits, from the last formula and the observation that for $\eta \in S(\mathbf{R})$, $: e^{\langle \cdot, \eta\rangle} : \varphi_{-\eta} \in (S)$ we conclude that the translation Φ_η of $\Phi \in (S)^*$, defined by

$$\langle \Phi_\eta, \varphi\rangle := \langle \Phi, : e^{\langle \cdot, \eta\rangle} : \varphi_{-\eta}\rangle \tag{2.24}$$

belongs to $(S)^*$. In fact, if $\Phi \in (S)_{-p}$, then

$$|\langle \Phi_\eta, \varphi\rangle| \leq \|\Phi\|_{2,-p}\| : e^{\langle \cdot, \eta\rangle} : \varphi_{-\eta}\|_{2,p} \\ \leq \|\Phi\|_{2,-p}\| : e^{\langle \cdot, \eta\rangle} : \|_{2,p+1}\|\varphi_{-\eta}\|_{2,p+1} \\ \leq \text{const.}\|\varphi\|_{2,q}$$

for some $q > p + 1$.

Corollary 2.8. *Let $\varphi \in (S)$, $y \in S'(\mathbf{R})$, and $\Phi \in (S)^*$. Then*

$$\langle \Phi, \varphi_y \rangle = \langle \Phi : (: e^{\langle \cdot, y \rangle} :), \varphi \rangle. \tag{2.25}$$

Proof. *Assume that φ and Φ are given by $(f^{(n)}, n \in \mathbf{N}_0)$ and $(T^{(n)}, n \in \mathbf{N}_0)$ respectively. Then*

$$\begin{aligned}\langle \Phi, \varphi_y \rangle &= \sum_{n=0}^{\infty} \sum_{k=0}^{n} \binom{n}{k} (n-k)! \langle T^{(n-k)}, f_y^{(n,k)} \rangle \\ &= \sum_{n=0}^{\infty} n! \sum_{k=0}^{n} \langle T^{(n-k)} \hat{\otimes} (\frac{1}{k!} y^{\otimes k}), f^{(n)} \rangle \\ &= \langle \Phi : (: e^{\langle \cdot, y \rangle} :), \varphi \rangle. \quad \square\end{aligned}$$

For the subspace $\mathcal{D}^*$ of $(S)^*$ we can relax the condition on the direction of the translation.

Proposition 2.9. *Let $\Phi \in \mathcal{D}^*$, $y \in L^2(\mathbf{R})$. Define*

$$\langle \Phi_y, \varphi \rangle := \langle \Phi, : e^{\langle \cdot, y \rangle} : \varphi_{-y} \rangle. \tag{2.26}$$

Then $\Phi_y \in (S)^*$. If $\Phi \in (L^2)$, then $\Phi_y(x) = \Phi(x+y)$ for μ-a.e. $x \in S'(\mathbf{R})$.

Proof. Assume that $\Phi \in \mathcal{D}_{p,-r}$, the dual of $\mathcal{D}_{q,r}$ with $q^{-1} + p^{-1} = 1$. Since $(S) \subset \mathcal{D}$ one has for some q', q'' with $q'^{-1} + q''^{-1} = q^{-1}$

$$\begin{aligned}|\langle \Phi, : e^{\langle \cdot, y \rangle} : \varphi_{-y} \rangle| &\le |\Phi|_{p,-r} |: e^{\langle \cdot, y \rangle} : \varphi_{-y}|_{q,r} \\ &\le |\Phi|_{p,-r} |: e^{\langle \cdot, y \rangle} :|_{q',r} |\varphi_y|_{q'',r}\end{aligned}$$

due to Meyer's inequalities [Me 83]. By inequality (2.16) and the observation that $: e^{\langle \cdot, y \rangle} : \in \mathcal{D}$ we find

$$|\langle \Phi, : e^{\langle \cdot, y \rangle} : \varphi_y \rangle| \le \text{const.} \|\varphi\|_{2,w}$$

for some $w \in \mathbf{N}_0$. $\square$

We conclude this section with a discussion of scaling transformations on (S). Let $\varphi \in (S)$ and for $\lambda \in \mathbf{R}$ put

$$(\varphi^{(\lambda)})^{\tilde{}}(x) := \tilde{\varphi}(\lambda x). \tag{2.27}$$

Clearly, $\varphi^{(\lambda)}$ denotes the μ-class of $(\varphi^{(\lambda)})^{\tilde{}}$.

Lemma 2.10. *Let $\lambda \in \mathbf{R}$, $\lambda \neq 0$, $n \in \mathbf{N}$, $x \in \mathcal{S}'(\mathbf{R})$. Then*

$$:(\lambda x)^{\otimes n}: \; = \lambda^n \sum_{k=0}^{[\frac{n}{2}]} \binom{n}{2k} (2k-1)!!(1-\lambda^{-2})^k \times \tag{2.28}$$
$$\times :x^{\otimes(n-2k)}: \hat{\otimes} Tr^{\otimes k}$$

$$= \lambda^n \sum_{k=0}^{[\frac{n}{2}]} \frac{n!}{k!(n-2k)!2^k} (1-\lambda^{-2})^k \times \tag{2.29}$$
$$\times :x^{\otimes(n-2k)}: \hat{\otimes} Tr^{\otimes K}$$

where $Tr \equiv \int dt \delta_t^{\otimes 2} \in \mathcal{S}'\widehat{(\mathbf{R}^2)}$.

Proof. In view of equation (2.5), we only need to show the following formula for Hermite polynomials

$$:(\lambda u)^n:_{\sigma^2} = \lambda^n \sum_{k=0}^{[\frac{n}{2}]} \binom{n}{2k} (2k-1)!!((1-\lambda^{-2})\sigma^2)^k :u^{n-2k}:_{\sigma^2} .$$

However, one can show this relation easily by a straightforward computation, using (2.4) and the fact that

$$:u^n:_{\sigma^2} \big|_{u=0} = \begin{cases} 0 & \text{if } n \text{ is odd} \\ (-1)^k (2k-1)!! & \text{if } n = 2k \end{cases} .$$

The details are left to the interested reader. □

Theorem 2.11. For every $\lambda \in \mathbf{R}$ the mapping $\varphi \mapsto \varphi^{(\lambda)}$ is continuous from $(\mathcal{S})$ into itself.

Proof. If $\lambda = 0$, $\varphi^{(\lambda)} = \varphi(0) = \langle \delta_0, \varphi \rangle$, so that $|\varphi(0)| \leq \|\delta_0\|_{2,-p} \|\varphi\|_{2,p}$, for any $p \in \mathbf{N}$, and the statement is trivial. Let $\lambda \in \mathbf{R} \setminus \{0\}$ and let $\varphi \in (\mathcal{S})$ be given by the sequence $(f^{(n)}, n \in \mathbf{N}_0)$. Identify $f^{(n)}$ with its representant in $\mathcal{S}(\mathbf{R}^n)$. From Lemma 2.10 we know that

$$\varphi^{(\lambda)}(x) = \sum_{n=0}^{\infty} \lambda^n \sum_{k=0}^{[\frac{n}{2}]} \binom{n}{2k} (2k-1)!!(1-\lambda^{-2})^k \times$$
$$\times \langle :x^{\otimes(n-2k)}:, \langle Tr^{\otimes k}, f^{(n)} \rangle \rangle$$

and therefore for any $r \in \mathbf{R}$ we have

$$\Gamma(A)^r \varphi^{(\lambda)}(x) = \sum_{n=0}^{\infty} \lambda^n \sum_{k=0}^{[\frac{n}{2}]} \binom{n}{2k} (2k-1)!!(1-\lambda^{-2})^k \times \\ \times \langle : x^{\otimes(n-2k)} :, (A^{\otimes(n-2k)})^r \langle Tr^{\otimes k}, f^{(n)} \rangle \rangle.$$

Now let $r \geq 0$, then we can estimate in the following way.

$$\|\Gamma(A)^r \varphi^{(\lambda)}\|_2 \leq \sum_{n=0}^{\infty} |\lambda|^n \sum_{k=0}^{[\frac{n}{2}]} \binom{n}{2k} (2k-1)!!|1-\lambda^{-2}|^k ((n-2k)!)^{1/2} \times \\ \times \|(A^{\otimes(n-2k)})^r \langle Tr^{\otimes k}, f^{(n)} \rangle\|_2$$

where the last norm is the one of $L^2(\mathbf{R}^{n-2k})$. Choose $p = r \vee 1$. Now we bound $(2k-1)!! \cdot ((n-2k)!)^{1/2}$ from above by const. $\cdot (n!)^{1/2}$ and use the estimate

$$\|(A^{\otimes(n-2k)})^r \langle Tr^{\otimes k}, f^{(n)} \rangle\|_2 \leq \|f^{(n)}\|_{2,p} \|A^{-p}\|_{H.S.}^{2k}$$

which is proved similarly as a corresponding estimate in the proof of Theorem 2.5. Here we used the notation $\|f^{(n)}\|_{2,p} := \|(A^{\otimes n})^p f^{(n)}\|_2$, the last norm being the one of $L^2(\mathbf{R}^n)$. Thus we have

$$\|\Gamma(A)^r \varphi^{(\lambda)}\|_2 \leq \text{const.} \sum_{n=0}^{\infty} (n!)^{\frac{1}{2}} |\lambda|^n \|f^{(n)}\|_{2,p} \times \\ \times \sum_{k=0}^{[\frac{n}{2}]} \binom{n}{2k} |1-\lambda^{-2}|^k \|A^{-p}\|_{H.S}^{2k} \\ \leq \text{const.} \sum_{n=0}^{\infty} (n!)^{\frac{1}{2}} |\lambda|^n \|f^{(n)}\|_{2,p} (1 + \|A^{-p}\|_{H.S.} |1-\lambda^{-2}|^{\frac{1}{2}})^n.$$

Choose $q > p$ large enough so that

$$2^{-(q-p)} |\lambda| (1 + \|A^{-p}\|_{H.S.} |1-\lambda^{-2}|^{1/2}) < 1.$$

Then we find by Schwarz' inequality

$$\|\Gamma(A)^r \varphi^{(\lambda)}\|_2 \leq \text{const.} (\sum_{n=0}^{\infty} n! \|f^{(n)}\|_{2,q}^2)^{1/2} \\ = \text{const.} \|\varphi\|_{2,q}.$$

It is clear that the statement of the Theorem follows from the last inequality.

Now we discuss some consequences concerning measures represented by elements in $(S)^*_+$. In the following let $\lambda \in \mathbf{R} \setminus \{0\}$. As before, if the measure ν on $(S'(\mathbf{R}), \mathcal{B})$ is represented by an element in $(S)^*_+$, we denote this element by $\frac{d\nu}{d\mu}$. Also, for $E \in \mathcal{B}$ we set $\nu^{(\lambda)}(E) := \nu(\lambda^{-1}E)$.

First we consider the white noise measure μ itself, i.e. $1 \in (S)^*$. Because of

$$\int \tilde{\varphi}(x) d\mu^{(\lambda)}(x) = \int \tilde{\varphi}(\lambda x)) d\mu(x)$$
$$= \langle 1, \varphi^{(\lambda)} \rangle.$$

it follows that for some $q \in \mathbf{N}_0$

$$\left| \int \varphi(x) d\mu^{(\lambda)}(x) \right| \leq \text{const.} \|\varphi\|_{2,q}$$

and therefore $\mu^{(\lambda)}$ is represented by an element $\frac{d\mu^{(\lambda)}}{d\mu} \in (S)^*$. The choice $\varphi(x) =: \exp\langle x, \xi\rangle :$, $\xi \in S(\mathbf{R})$, gives

$$S \frac{d\mu^{(\lambda)}}{d\mu}(\xi) = e^{-\frac{1}{2}(1-\lambda^2)\|\xi\|_2^2}. \tag{2.30}$$

More generally we can prove the following result.

Theorem 2.12. Assume that ν is a measure on $(S'(\mathbf{R}), \mathcal{B})$ so that $\frac{d\nu}{d\mu} \in (S)^*$. Then for all $\lambda \in \mathbf{R} \setminus \{0\}$, $\frac{d\nu^{(\lambda)}}{d\mu} \in (S)^*$. Moreover

$$S \frac{d\nu^{(\lambda)}}{d\mu}(\xi) = (S \frac{d\nu}{d\mu})(\lambda\xi) \cdot (S \frac{d\mu^{(\lambda)}}{d\mu})(\xi), \xi \in S(\mathbf{R}). \tag{2.31}$$

Proof. Let $\varphi \in (S)$. Then

$$\int \tilde{\varphi}(x) d\nu^{(\lambda)}(x) = \int (\varphi^{(\lambda)})^{\sim}(x) d\nu(x)$$
$$= \langle \frac{d\nu}{d\mu}, \varphi^{(\lambda)} \rangle.$$

Without loss of generality we may assume that $\frac{d\nu}{d\mu} \in (S)_{-p}$, for some $p \in \mathbf{N}_0$ and Theorem 2.11 implies the estimate

$$\left| \int \varphi(x) d\nu^{(\lambda)}(x) \right| \leq \|\frac{d\nu}{d\mu}\|_{2,-p} \|\varphi^{(\lambda)}\|_{2,p}$$
$$\leq \text{const.} \|\frac{d\nu}{d\mu}\|_{2,-p} \|\varphi\|_{2,q}.$$

for some $q > p$, $q \in \mathbf{N}$. Thus the first part of the statement of the theorem is proven. To show (2.31), we choose again $\varphi =: \exp\langle\cdot, \xi\rangle :$, $\xi \in S(\mathbf{R})$ and find immediately

$$\begin{aligned}\langle\frac{d\nu^{(\lambda)}}{d\mu}, : e^{\langle\cdot,\xi\rangle} :\rangle &= \langle\frac{d\nu}{d\mu}, e^{\lambda\langle x,\xi\rangle}\rangle \cdot e^{-\frac{1}{2}\|\xi\|_2^2} \\ &= \langle\frac{d\nu}{d\mu}, : e^{\lambda\langle\cdot,\xi\rangle} :\rangle \cdot e^{-\frac{1}{2}(1-\lambda^2)\|\xi\|_2^2}.\end{aligned}$$

For $\lambda \in \mathbf{R}\setminus\{0\}$ and $\Phi \in (S)^*$ given by the sequence $(T^{(n)}, n \in \mathbf{N}_0)$, let us set $\Phi_{(\lambda)}$ as being defined by the sequence of kernels $(\lambda^n T^{(n)}, n \in \mathbf{N}_0)$. It is easy to see that $\Phi_{(\lambda)} \in (S)^*$ and that the restriction of the mapping $\Phi \mapsto \Phi_{(\lambda)}$ to (S) acts continuously on (S). Also we have

$$S\Phi_{(\lambda)}(\xi) = S\Phi(\lambda\xi), \xi \in S(\mathbf{R}).$$

Then we can rewrite (2.31) as follows

$$\frac{d\nu^{(\lambda)}}{d\mu} = (\frac{d\nu}{d\mu})_{(\lambda)} : \frac{d\mu^{(\lambda)}}{d\mu}. \tag{2.32}$$

3. Differential operators on (S)

In this section we discuss Gâteaux derivatives D_y, $y \in S'(\mathbf{R})$, acting on elements of (S) and compare the operator D_y for the choice $y = \delta_t$, $t \in \mathbf{R}$, with the Hida derivative [Hi 75, KT 80, Ku 83]. We shall show that for every $y \in S'(\mathbf{R})$, D_y acts continuously on (S) and every $\varphi \in (S)$ is C^∞ with respect to D_y. Also we shall prove that for every $q > \frac{1}{4}$ $(S)_q$ is a subspace of the (L^2)-domain of the Hida derivative. Finally, Taylor's formula for the translation operator τ_y, $y \in S'(\mathbf{R})$, will be derived. This shows the 'analyticity' of the elements of (S).

First consider the algebra $\mathcal{P}$ of polynomials on $S'(\mathbf{R})$. $P \in \mathcal{P}$ is of the form

$$P(x) = \sum_n \sum_{\alpha\in\mathbf{N}^n} a_\alpha \langle x, \xi_{\alpha_1,n}\rangle \cdots \langle x, \xi_{\alpha_n,n}\rangle \tag{3.1}$$

with complex numbers a_α, $\xi_{\beta,n} \in S'(\mathbf{R})$, both sums in (3.1) having only a finite number of non-zero terms. P can be rewritten in terms of Wick-ordered monomials

$$P(x) = \sum_n \sum_{\alpha\in\mathbf{N}^n} b_\alpha : \langle x, \eta_{\alpha_1,n}\rangle \cdots \langle x, \eta_{\alpha_n,n}\rangle : \tag{3.2}$$

and therefore may be represented in the standard form as

$$P(x) = \sum_n \langle : x^{\otimes n} :, f^{(n)}\rangle \tag{3.3a}$$

$$f^{(n)} = \sum_\alpha b_\alpha \hat{\otimes}_{j=1}^n \eta_{\alpha_j,n}. \tag{3.3b}$$

Let F be a (real - or complex valued, say) pointwise defined functional on $\mathcal{S}'(\mathbf{R})$ and $y \in \mathcal{S}'(\mathbf{R})$. If for $x \in \mathcal{S}'(\mathbf{R})$ the function $s \mapsto F(x+sy)$ is differentiable at $s=0$, we say that F is *Gâteaux differentiable at x in direction y* and denote

$$D_y F(x) := \frac{d}{ds} F(x+sy)|_{s=0} \tag{3.4}$$

called the *Gâteaux derivative of F at x in direction y*. It is obvious that the functional $\langle \cdot, \xi \rangle$, $\xi \in \mathcal{S}(\mathbf{R})$, is everywhere Gâteaux differentiable in any direction of $\mathcal{S}'(\mathbf{R})$, so that the chain rule for D_y (cf. e.g. [Sch 69] implies that for all $y \in \mathcal{S}'(\mathbf{R})$, all $P \in \mathcal{P}$ are everywhere Gâteaux differentiable in direction y. Moreover it is easy to compute that P of the form (3.1) - (3.3) has Gâteaux derivative

$$D_y P(x) = \sum_n \sum_\alpha a_\alpha \sum_{k=1}^{n} \langle y, \xi_{\alpha_k,n} \rangle \times \\ \times \langle x, \xi_{\alpha_1,n} \rangle \cdots \langle x, \xi_{\alpha_k,n} \rangle \cdots \langle x, \xi_{\alpha_n,n} \rangle \tag{3.5}$$

$$= \sum_n \sum_\alpha b_\alpha \sum_{k=1}^{n} \langle y, \eta_{\alpha_k,n} \rangle \times \\ \times : \langle x, \eta_{\alpha_1,n} \rangle \cdots \langle x, \eta_{\alpha_k,n} \rangle \cdots \langle x, \eta_{\alpha_n,n} \rangle : \tag{3.6}$$

$$= \sum_n n \langle : x^{\otimes n} :, \langle y, f^{(n)} \rangle \rangle \tag{3.7}$$

where in the last equation $\langle y, f^{(n)} \rangle$ means the pairing of y with $f^{(n)}$ in its first argument (which is well-defined, since $f^{(n)}$ is the sum of symmetric tensor products of Schwartz space test functions), so that $\langle y, F^{(n)} \rangle \in \mathcal{S}(\widehat{\mathbf{R}^{(n-1)}})$.

For the special choice $y = \delta_t$, $t \in \mathbf{R}$, we obtain

$$D_{\delta_t} = \sum_n \langle : x^{\otimes (n-1)} :, n f^{(n)}(t, \cdot) \rangle. \tag{3.8}$$

Let us identify $\mathcal{P}$ with its class in (L^2). For $t \in \mathbf{R}$, the Hida derivative $\partial_t \varphi$ of $\varphi \in (L^2)$ is defined (whenever it exists) by

$$\partial_t \varphi(x) := \mathcal{S}^{-1} \frac{\delta}{\delta \xi(t)} \mathcal{S} \varphi(x) \tag{3.9}$$

$\frac{\delta}{\delta \xi(t)}$ denoting the Fréchet functional derivative, acting on $\mathcal{S}\varphi$ which is considered as a functional of $\xi \in \mathcal{S}(\mathbf{R})$. (Here we have given the definition of ∂_t of Kubo and Takenaka [KT 80], which is equivalent to the definitions of Hida (e.g. [Hi 75]) and of Kuo [Ku 83]). For $P \in \mathcal{P}$ of the form above it is immediate to compute

$$\partial_t P(x) = D_{\delta_t} P(x), (\mu - \text{a.e.}) x \in \mathcal{S}'(\mathbf{R}), t \in \mathbf{R} \tag{3.10}$$

since

$$SP(\xi) = \sum_n \int_{\mathbf{R}^n} f^{(n)}(u)\xi^{\otimes n}(u)du. \tag{3.11}$$

Clearly, in order that (3.10) makes sense for every $t \in \mathbf{R}$ and every $x \in S'(\mathbf{R})$, one just picks the pointwise defined, continuous version of P.

The following formula will be useful later.

Lemma 3.1. *For $h \in L^2(\mathbf{R})$, $P \in \mathcal{P}$ we have*

$$D_h P = \int dt\ h(t)\ \partial_t P. \tag{3.12}$$

Here we have identified D_{δ_t} with ∂_t on $\mathcal{P}$ in accordance with (3.10).

Proof. From (3.5) or (3.6) we conclude that $h \mapsto D_h P(x)$ is a continuous linear functional on $L^2(\mathbf{R})$ for every $P \in \mathcal{P}$ and every $x \in S'(\mathbf{R})$. Consequently it is of the form

$$D_h P(x) = \int_{\mathbf{R}} dt\ h(t)\ (\nabla P(x))(t) \tag{3.13}$$

for some $\nabla P(x) \in L^2(\mathbf{R})$. Comparison of (3.6) with (3.8) (using (3.3b)) shows that $(\nabla P(x))(t) = \partial_t P(x)$. □

In order to extend the derivatives to (S) we first prove the following estimate.

Lemma 3.2. *For all p, $q \in \mathbf{N}_0$, $h \in L^2(\mathbf{R})$ and $\varphi \in \mathcal{P}$*

$$\|D_h\varphi\|_{2,p} \le C_{p,q}\|h\|_{2,-q}\|N^{1/2}2^{-(p\vee q-p)N}\varphi\|_{2,p\vee q} \tag{3.14}$$

where $C_{p,q} = 2^{(p\vee q-p)}$ and $p \vee q = \max\{p, q\}$.

Proof. Representing φ as in (3.3) we know by (3.8) and (3.13) that $D_h\varphi$ corresponds to the sequence $(n\int h(t)\ f^{(n)}(t,\cdot)\ dt, n \in \mathbf{N})$ of kernel functions. Therefore*

$$\begin{aligned}\|D_h\varphi\|_{2,p}^2 &= \sum_{n=1}^{\infty}(n-1)!n^2\|(A^{\otimes(n-1)})^p \int h(t)\ f^{(n)}(t,\cdot)\ dt\|_2^2 \\ &= \sum_{n=1}^{\infty} n!n\|(A^{\otimes(n-1)})^p \int (A^{-q}h)(t)\ (A_1^q f)^{(n)}(t,\cdot)\ dt\|_2^2\end{aligned}$$

* $A_1 f^{(n)}$ means that the differential operator A acts in the first variable of $f^{(n)}$

$$\leq \sum_{n=1}^{\infty} n!n\|A^{-q}h\|_2^2 \|A^{\otimes(n-1)p} \otimes A_1^q f^{(n)}\|_2^2$$

$$\leq \|A^{-q}h\|_2^2 \sum_{n=1}^{\infty} n!nC_{p,q}^{-2(n-1)} \|f^{(n)}\|_{2,p\vee q}^2$$

$$= C_{p,q}^2 \|h\|_{2,-q}^2 \|N^{1/2} 2^{-(p\vee q-p)N} \varphi\|_{2,p\vee q}^2. \quad \Box$$

Corollary 3.3. *For all $h \in S'(\mathbf{R})$, $\varphi \in P$, $p \in \mathbf{N}_0$*

$$\|D_h\varphi\|_{2,p} \leq \|h\|_2 \|N^{1/2}\varphi\|_{2,p} \tag{3.15}$$

$$\leq \|h\|_2 \|\varphi\|_{2,p+1} \tag{3.16}$$

Clearly, (3.16) allows to extend D_h to (S):

Proposition 3.4. *For all $h \in L^2(\mathbf{R})$, D_h extends to a continuous linear map from (S) into itself, such that for all p, $q \in \mathbf{N}_0$ (3.14) holds for all $\varphi \in (S)$.*

Since every element $\varphi \in (S)$ has a unique, strongly continuous version $\tilde{\varphi}$, the question arises whether

$$(D_h\varphi)^{\sim}(x) = D_h\tilde{\varphi}(x), x \in S'(\mathbf{R}). \tag{3.17}$$

In fact, it is not hard to see that $(\varphi^{(n)})^{\sim}$, $n \in \mathbf{N}$, is everywhere C^∞ with respect to D_y, $y \in S'(\mathbf{R})$, due to the nuclear theorem. Moreover, $\tilde{\varphi}$ is a sum of the $(\varphi^{(n)})^{\sim}$'s which converges pointwwise, uniformly on the strongly bounded sets (cf. [KY 87] and the arguments in [AHP 89b, Appendix]). Using estimates quite similar to those in the proof of the existence of $\tilde{\varphi}$ in [KY 87], we find that the sum of the expressions $(\frac{d}{ds}\varphi^{(n)}(x+sy))^{\sim}$ converges for all x, $y \in S'(\mathbf{R})$, uniformly in s on bounded subsets of $\mathbf{R}$. Therefore we conclude that the pointwise defined version $\tilde{\varphi}$ of $\varphi \in (S)$ is everywhere C^∞ in any direction $y \in S'(\mathbf{R})$. Thus the right hand side of (3.17) makes sense.

Clearly, for polynomials φ (3.17) holds. Since the evaluation map $\varphi \mapsto \tilde{\varphi}(x)$, $x \in S'(\mathbf{R})$, belongs to $(S)^*$ and φ is the (S) - limit of polynomials, we see that (3.17) holds for all $\varphi \in (S)$.

Next we extend to derivatives in direction of general $y \in (S)'(\mathbf{R})$.

Theorem 3.5. *For all $y \in S'(\mathbf{R})$, D_y extends from P to a continuous linear mapping from (S) into itself. Moreover, if $y \in S_{-q}(\mathbf{R})$, then*

$$\|D_y\varphi\|_{2,p} \leq C_{p,q}\|y\|_{2,-q}\|N^{1/2}2^{-(p\vee q-p)N}\varphi\|_{2,p\vee q}. \tag{3.18}$$

Also, for all $x \in S'(\mathbf{R})$

$$(D_y\varphi)^{\sim}(x) = D_y\tilde{\varphi}(x). \tag{3.19}$$

Proof. *Let $y \in S_{-q}(\mathbf{R})$, $q \in \mathbf{N}_0$. There is a sequence $(y_n, n \in \mathbf{N})$ in $L^2(\mathbf{R})$, such that $y_n \to y$ in $S_{-q}(\mathbf{R})$. From Lemma 3.2 we conclude that $(D_{y_n}, n \in \mathbf{N})$ is a Cauchy in the Banach space of bounded linear mappings from $(S)_{p\vee q+1}$ into $(S)_p$, $p \in \mathbf{N}_0$, equipped with the uniform topology. Therefore, it has a limit which we also denote by D_y. Due to (3.14) we have the bound (3.18) for D_y and every $\varphi \in (S)$. Finally (3.19) follows, because it is true for $y \in L^2(\mathbf{R})$ and since $D_{y_n}\varphi \to D_y\varphi$ in (S), so that the remark that the evaluation map belongs to $(S)^*$ completes the proof.* □

In particular Theorem 3.5 shows that every $\varphi \in (S)$ is C^∞ (in (S)) under D_y for all $y \in S'(\mathbf{R})$.

Corollary 3.6. *For every $\varphi \in (S)$, the mapping $y \mapsto D_y\varphi$ is continuous with respect to the strong topology on $S'(\mathbf{R})$.*

Proof. Clearly we have for every $\varphi \in (S)$ and $y_1, y_2 \in S'(\mathbf{R})$ that $D_{y_1}\varphi - D_{y_2}\varphi = D_{y_1-y_2}\varphi$. Therefore inequality (3.18) gives the desired continuity with respect to the inductive limit topology on $S'(\mathbf{R})$. However the latter and the strong topology on $S'(\mathbf{R})$ coincide, cf. [AHP 89c, Appendix]. □

Consider now the Hida derivative ∂_t, $t \in \mathbf{R}$. We can localize Theorem 3.5 as follows.

Theorem 3.7. *For all $t \in \mathbf{R}$, $p \geq 0$ and $q > \frac{1}{4}$, ∂_t is a continuous mapping from $(S)_{p+q}$ into $(S)_p$. In particular $(S)_q$ belongs to the (L^2) - domain of ∂_t.*

Proof. In view of (3.18) it suffices to show that the Dirac distribution δ_t, $t \in \mathbf{R}$, belongs to $S_{-q}(\mathbf{R})$, for $q > \frac{1}{4}$, i.e. that $A^{-q}\delta_t \in L^2(\mathbf{R})$. This can be seen as follows. Write

$$\langle \delta_t, \cdot \rangle = \sum_{k=0}^{\infty} e_k(t)\ (e_k, \cdot) \tag{3.20}$$

where $(e_k, k \in \mathbf{N}_0)$ is the CONS in $L^2(\mathbf{R})$ made up from the Hermite functions. (The right hand side of (3.20) converges weakly in $S'(\mathbf{R})$.) Then

$$\|A^{-q}\delta_t\|_2^2 = 2^{-q}\sum_{k=0}^{\infty}(k+1)^{-2q}e_k(t)^2.$$

Since

$$\sup_{t\in\mathbf{R}} |e_k(t)| = O(k^{-1/4})$$

(cf. e.g. [HPh 57, 21.3.2]), our claim follows. □

We conclude this section by proving Taylor's formula on (S).

Theorem 3.8. *Let $y \in S'(\mathbf{R})$, $\varphi \in (S)$. Then*

$$\tau_y \varphi = (\exp\ D_y)\varphi \tag{3.21}$$

i.e.

$$\sum_{n=0}^{N} \frac{1}{n!} D_y^n \varphi \to \tau_y \varphi, N \to \infty$$

in (S).

Proof. From Theorem 2.5 and its proof we know that for $\varphi \in (S)$, corresponding to the sequence $(f^{(n)}, n \in \mathbf{N}_0)$ we have (μ - a.e.)

$$\tau_y \varphi(x) = \sum_{n=0}^{\infty} \sum_{m=0}^{n} \binom{n}{m} \langle : x^{\otimes m} :, \langle y^{\otimes(n-m)}, f^{(n)} \rangle\rangle \tag{3.22}$$

the last sum being convergent in (S).

On the other hand it follows easily from (3.7) and the fact that D_y is a closed operator on (S) that

$$D_y^m \varphi(x) = \sum_n \frac{n!}{(n-m)!} \langle : x^{\otimes m} :, \langle y^{\otimes(n-m)}, f^{(n)} \rangle\rangle \tag{3.23}$$

(with the convention $(n-m)!^{-1} = 0$ if $m > n$). Thus

$$\sum_{m=0}^{N} \frac{1}{m!} D_y^m \varphi(x) = \sum_{n=0}^{\infty} \sum_{m=0}^{N\wedge n} \binom{n}{m} \langle : x^{\otimes m} :, \langle y^{\otimes(n-m)}, f^{(n)} \rangle\rangle$$

and

$$(\tau_y - \sum_{m=0}^{N} \frac{1}{m!} D_y^m)\varphi = \sum_{n=N+1}^{\infty} \sum_{m=0}^{N} \binom{n}{m} \langle : x^{\otimes m} :, \langle y^{\otimes(n-m)}, f^{(n)} \rangle\rangle.$$

As the estimation in the proof of Theorem 2.5 shows, the last sum is the Cauchy remainder of a sum convergent in (S) and therefore it vanishes in (S) as N tends to infinity. □

Because the evaluation map belongs to $(S)^*$, we have

Corollary 3.9. *For every $\varphi \in (S)$, x, $y \in S'(\mathbf{R})$*

$$\tilde{\varphi}(x+y) = \sum_{n=0}^{\infty} \frac{1}{n!} D_y^n \tilde{\varphi}(x) \tag{3.24}$$

$$= \sum_{n=0}^{\infty} \frac{1}{n!} (D_y^n \varphi)^{\sim}(x). \tag{3.25}$$

According to Corollary 3.8 we may say that the elements of (S) are *analytic*.

4. Convolutions of measures on $S'(\mathbf{R})$

Consider the functional

$$\Xi(\xi) := \exp(\frac{1}{2}\|\xi\|^2_{L^2(\mathbf{R})}), \; \xi \in S(\mathbf{R}) \tag{4.1}$$

on $S(\mathbf{R})$. We assert the reader that Ξ is the S-transform of an element in $(S)^*$, which we denote by Φ_0. In fact, by (2.30) one has $\Phi_0 = d\mu^{(\sqrt{2})}/d\mu$.

Now let ν_1, ν_2 be two measures on $(S'(\mathbf{R}), \mathcal{B})$, with $\frac{d\nu_j}{d\mu} \in (S)^*_+$, $j = 1,2$. Define an element Φ of $(S)^*$ by

$$\Phi := \frac{d\nu_1}{d\mu} : \frac{d\nu_2}{d\mu} : \Phi_0. \tag{4.2}$$

Compute the S-transform of Φ at $\lambda\xi$, $\xi \in S(\mathbf{R})$, $\lambda \in \mathbf{R}$:

$$\begin{aligned}(S\Phi)(\lambda\xi) &= (S\frac{d\nu_1}{d\mu})(\lambda\xi) \cdot (S\frac{d\nu_2}{d\mu})(\lambda\xi) \cdot \exp(\frac{1}{2}\lambda^2\|\xi\|_2^2) \\ &= \int e^{\lambda}\langle x, \xi\rangle d\nu_1(x) \cdot \int e^{\lambda\langle x,\xi\rangle} d\nu_2(x) \times \\ &\quad \times \exp(-\frac{1}{2}\lambda^2\|\xi\|_2^2) \\ &= \int e^{\lambda\langle x,\xi\rangle} d(\nu_1 * \nu_2)(x) \cdot \exp(-\frac{1}{2}\lambda^2\|\xi\|_2^2)\end{aligned}$$

where $\nu_1 * \nu_2$ denotes the convolution of the measure ν_1, ν_2.

It can be shown that $\lambda \mapsto S\Phi(\lambda\xi)$, $\xi \in S(\mathbf{R})$, $\Phi \in (S)^*$, has an analytic continuation to the whole complex plane (cf. e.g. [HKP]). Moreover, we have the formula

$$\langle \Psi, e^{i\langle\cdot,\xi\rangle}\rangle = \exp(-\frac{1}{2}\|\xi\|_2^2) S\Psi(i\xi) \tag{4.3}$$

for $\psi \in (S)^*$. Thus we find

$$\langle \Phi, e^{i\langle\cdot,\xi\rangle}\rangle = \int e^{i\langle x,\xi\rangle} d(\nu_1 * \nu_2)(x). \tag{4.4}$$

Because $\nu_1 * \nu_2$ is positive, it follows that $\xi \mapsto \langle \Phi, \exp(i\langle\cdot,\xi\rangle)\rangle$ is positive definite. From the results in [Yo 87] we conclude that Φ is positive and represented by a measure, which necessarily is equal to $\nu_1 * \nu_2$. Therefore we have derived the following result.

Theorem 4.1. *Let ν_1, ν_2 be two (positive) measures on $(S'(\mathbf{R}), \mathcal{B})$, such that $\frac{d\nu_j}{d\mu} \in (S)^*$, $j = 1,2$. Then the convolution $\nu_1 * \nu_2$ is also represented by an element $\frac{d(\nu_1 * \nu_2)}{d\mu}$ in $(S)^*_+$ and the following formula holds*

$$\frac{d(\nu_1 * \nu_2)}{d\mu} = \frac{d\nu_1}{d\mu} : \frac{d\nu_2}{d\mu} : \Phi_0 \tag{4.5}$$

where Φ_0 is the element in $(S)^*$ whose S-transform is given by (4.1).

References

[AHP 89a] Albeverio, S., Hida, T., Potthoff, J., Streit, L.: The vacuum of the Høgh-Krohn model as a generalized white noise functional; *Phys. Lett. B* **217** (1989) 511-514.

[AHP 89b] Albeverio, S., Hida, T., Potthoff, J., Röckner, M., Streit, L.: Dirichlet forms in terms of white noise analysis I – Construction and QFT examples; *Reviews in Math. Phys.* **1** (1990) 291-312.

[AHP 89c] Albeverio, S., Hida, T., Potthoff, J., Röckner, M., Streit, L.: Dirichlet forms in terms of white noise analysis II – Closability and diffusion processes; *Reviews in Math. Phys.* **1** (1990) 313-323.

[AP 88] Asch, J., Potthoff, J.: Itô's lemma without non-anticipatory condition; *Preprint* (1988) to appear in *Probab. Th. Rel. Fields.*

[Ch 88] Chow, P.L.: Stochastic differential equaitons in infinite dimensions; in: *Mathematics + Physics,* **Vol.3,** ed. by J. Potthoff and L. Streit; Singapore: World Scientific (1988).

[GV 84] Gel'fand, I.M,. Vilenkin, N. Ya.: *Generalized functions,* **Vol. 4**; New York, London: Academic Press (1964).

[Hi 75] Hida, T.: *Analysis of Brownian functionals;* Carleton Mathematical Lecture Notes no. **13** (1975).

[Hi 80] Hida, T.: *Brownian motion;* Berlin, Heidelberg, New York: Springer (1980).

[HKP] Hida, T., Kuo, H,-H., Potthoff, J., Streit, L.: *White noise: an infinite dimensional calculus;* monograph in preparation.

[HPS 88a] Hida, T., Potthoff, J., Streit, L.: Dirichlet forms and white noise analysis; *Comm. Math. Phys.* **116** (1988) 235-245.

[HPS 88b] Hida, T., Potthoff, J., Streit, L.: White noise analysis and applications; *BiBos Preprint* (1988), *Mathematics + Physics*, **Vol. 3**; ed. by L. Streit, Singapore: World Scientific (1988) 143-178.

[HS 83] Hida, T., Streit, L.: Generalized Brownian functionals and the Feynman integral; *Stoch. Proc. Appl.* **16** (1983) 55-69.

[HPh 57] Hille, E., Phillips, R.S.: *Functional analysis and semigroups* Amer. Math. Soc. Colloq. Pub. Vol. XXI (1957).

[KU 89] Korezlioglu, H., Ustunel, A.S: New class of distributions on the Wiener space; *Preprint* (1989).

[Kb 83] Kubo, I.: Itô's formula for generalized Brownian functionals; in: *Theory and applications of random fields,* ed. by G. Kallianpur; Berlin, Heidelberg New York: Springer (1983).

[KT 83] Kubo, I., Takenaka, S.: Calculus on Gaussian white noise; **I-IV** *Proc. Japan Acad.* **56** (1980) 376-380, 411-416, **57** (1981) 433-437, **58** (1982) 186-189.

[KY 87] Kubo, I., Yokoi, Y.: A remark on the space of testing random variables in the white noise calculus; *Nagoya Math. J.* **115** (1989) 139-149.

[Ku 83] Kuo, H.-H.: Brownian functionals and applications; *Acta Appl. Math.* **1** (1983) 175-188.

[KS 87] Kuo, H.-H., Shieh, N.-R.: A generalized Itô's formula for multidimensional Borwnian motions and its applications; *Chinese J. Math.* **15** (1987) 163-174.

[Me 83] Meyer, P.A.: Quelques résultats analytiques sur le semigroupe d'Orsterin-Uhlenbeck en dimension infinie; in: *Theory and application of random fields* ed. by G. Kallianpur, Berlin, Heidelberg, New York: Springer (1983).

[MY 87] Meyer, P.A., Yan, J.-A.: A propos des distributions sur l'espace de Wiener; in *Séminaire de Probabilités* **XXI**, ed by J. Azéma and M. Yor, Berlin, Heidelberg, New York: Springer (1987).

[MY 88] Meyer, P.A., Yan, J.-A.: Distributions sur l'espace de Wiener (suite); *Preprint* (1988), to appear in *Séminaire de Probabilités* **XXIII**.

[Ne 73] Nelson, E.: Probability theory and Euclidean quantum field theory; in *Constructive quantum field theory* ed. by G. Velo, A. Wightman: Berlin, Heidelberg, New York: Springer (1973).

[NZ 88] Nualart, D., Zakai, M.: Generalized Brownian functionals and the solution to a stochastic partial differential equation; *Preprint* (1988), to appear in *J. Funct. Anal.*.

[Po 87] Potthoff, J.: On positive generalized functionals; *J. Funct. Anal.* **74** (1987) 81-95.

[Po 88a] Plotthoff, J.: Littlewood-Paley theory on Gaussian spaces; *Nagoya Math. J.* **109** (1988) 47-61.

[Po 88b] Potthoff, J.: On Meyer's equivalence; *Nagoya Math. J.* **111** (1988) 99-109.

[Sch 69] Schwartz, J.: *Nonlinear functional analysis;* New York, London, Paris: Gordon and Breach (1969).

[Si 74] Simon, B.: *The $P(\varphi)_2$ Euclidean (quantum) field theory;* Princeton: Princeton University Press (1974).

[Su 85] Sugita, H.: Sobolev spaces of Wiener functionals and Malliavin's calculus; *J. Math. Kyoto Univ.* **25** (1985) 717-725.

[Tr 67] Trèves, F.: *Topological vector spaces, distributions and kernels;* New York, London: Academic Press (1967).

[Us 87] Ustunel, A, S.: Representation of the distributions on Wiener space and stochastic calculus of variations; *J. Funct. Anal.* **70** (1987) 126-134.

[Wa 83] Watanabe, S.: Maliavin's calculus in terms of generalized Wiener functionals; in *Theory and application of random fields;* ed. by G. Kallianpur, Berlin, Heidelberg, New York: Springer (1983).

[Yo 87] Yokoi, Y.: Positive generalized functionals; *Preprint* (1987), to appear in *Hiroshima Math. J.*

Tail approximations for maxima of random fields

David O. Siegmund

Abstract. The method for approximating the tail of the distribution of the maximum of a random field first developed by Woodroofe (1976, 1978) and extended to the multidimensional case by Siegmund (1988) is applied to approximate the significance level of the likelihood ratio test for no change–point against the alternative of at most two changes in the mean of a sequence of normal observations with common variance.

1. Introduction

Although the number of processes $X(t), t \in T$, for which one can evaluate exactly $\Pr\{\max X(t) > b\}$ is very small, the last twenty years have seen a variety of methods developed to obtain approximations to such probabilities, especially asymptotic approximations as $b \to \infty$ and hence the probability converges to zero.

The number of methods is quite large when the indexing set T is one–dimensional although admittedly even there different methods are more or less successful in dealing with different problems. When T is several dimensional, the number of available methods shrinks considerably. Historically the first of these is the asymptotic inclusion–exclusion method developed by Pickands (1969) in one dimension and exteded to $d > 1$ dimensions independently by Qualls and Watanabe (1973) and Bickel and Rosenblatt (1973). However, the resulting approximation involves the maximum of a derived random field and in general cannot be explicitly evaluated. Hogan and Siegmund (1986) observed that this method does yield explicitly, easily computed approximations in many special cases. More recently, a beautiful heuristic method for obtaining approximations in a number of problems has been developed by Aldous (1989). The structure of the underlying random field which is important for obtaining explicit approximations is that locally it behaves like a superposition of independent random walks. In Aldous (1989) terminology this is the "uncorrelated orthogonal increments property."

The subject of this paper is a third method, which seems to work particularly well in a variety of problems arising in statistics. The method was pioneered by Woodroofe (1976, 1978) in one dimensional problems of sequential analysis. Siegmund (1988) discussed a number of examples where the method can be adapted for random fields, particularly those arising in change–point problems. (See also Kim and Siegmund (1989).) This paper is concerned with a related example requiring some new techniques, which seem related to what Aldous (1989) calls the marked clump technique.

Probability Theory

Section 2 contains a description of a statistical problem and a statement of the main results. A proof is discussed informally in Section 3. Some supporting renewal theoretic lemmas appear in Section 4. Several technical calculations which appear in all boundary crossing problems have been omitted.

2. Tests to detect change–points

Let $x_n\,(n = 1, 2, ..., m)$ be independent and normally distributed. Assume there exist $1 \leq j_1 < j_2 \leq m$ such that x_n is $N(\mu_1, \sigma^2)$ for $n \leq j_1$, $N(\mu_2, \sigma^2)$ for $j_1 < n \leq j_2$, and $N(\mu_3, \sigma^2)$ for $j_2 < n \leq m$. The mean values μ_1, μ_2, μ_3 and the change–points j_1, j_2 are all unknown. Usually σ^2 is also unknown, but for simplicity it is assumed known and without loss of generality equal to one for most of this paper. We shall be concerned with testing the hypothesis of no change–point, $H_0 : \mu_1 = \mu_2 = \mu_3$, against the alternative that at least two of the μ_i are different. There is a substantial literature on the simpler model of at most one change–point, the special case in which it is known that $\mu_2 = \mu_3$ or equivalently $j_2 = m$. See, for example, James, James, and Siegmund (1987).

The likelihood ratio statistic for testing H_0 is as follows. Let $S_n = x_1 + \ldots + x_n$ and for $0 \leq i \leq j \leq m$ put $\bar{x}_{i,j} = (S_j - S_i)/(j - i)(\bar{x}_{i,i} = 0)$. Let

$$(1) \quad Q_{i,j} = i(\bar{x}_{0,i} - \bar{x}_{0,m})^2 + (j - i)(\bar{x}_{i,j} - \bar{x}_{0,m})^2 + (m - j)(\bar{x}_{j,m} - \bar{x}_{0,m})^2.$$

The (log) likelihood ratio statistic for testing H_0 is

$$(2) \qquad Q = \max_{1 \leq i < j \leq m} Q_{i,j}.$$

If the only possible change–points were $j_1 = i$, $j_2 = j$, we would be faced with the standard problem of testing the equality of means of three normal populations, and $Q_{i,j}$ would be the (log) likelihood ratio statistic. Under H_0 the distribution of $Q_{i,j}$ is χ^2 with two degrees of freedom. Since in fact j_1 and j_2 are unknown parameters, the (log) likelihood ratio statistic is $Q = \max Q_{i,j}$. The significance level of the likelihood ratio test is

$$(3) \qquad P\{Q^{1/2} > b\},$$

where P denotes probability under H_0. The principal results of this paper are approximations for (3) and for the significance level of the likelihood ratio test when σ^2 is unknown, which are given below in Theorems (9) and (10).

It will be convenient to introduce the following notation : $\varphi(x) = (2\pi)^{-1/2}\exp(-\frac{1}{2}x^2)$, $\Phi(x) = \int_{-\infty}^{x} \varphi(z)dz$, and

$$(4) \qquad \nu(x) = 2x^{-2} \exp\left\{-2\sum_{1}^{\infty} n^{-1}\Phi(-\frac{1}{2}xn^{1/2})\right\} \quad (x > 0).$$

A useful approximation for the function ν is

$$\nu(x) = \exp(-Px) + o(x^2) \quad (x \to 0), \tag{5}$$

where $\rho \cong .583$ (cf. Siegmund (1985), Proposition 10.37).

For the simpler alternative of at most one change the likelihood ratio test statistic is

$$\max_{1 \le k < m} \{(S_k - kS_m/m)^2/[k(1-k/m)]\}.$$

It is known (Siegmund (1985), Theorem 11.30) that if $m \to \infty$, $b \to \infty$, and $b/m^{1/2} \to b_0 > 0$, then under H_0 for any $0 \le t_0 < t_1 \le 1$,

$$P\Big\{\max_{mt_0 < k < mt_1} [|S_k - kS_m/m|/\{k(1-k/m)\}^{1/2}] > b\Big\} \sim b\varphi(b)\int_{t_0}^{t_1} [t(1-t)]^{-1}\nu\{b_0/(1-t)\}]^{1/2}\}dt. \tag{6}$$

The special function ν defined in (4) appears on the right hand side of (6) because the process $S_k - kS_m/m$ jumps over the curved boundary $b[k(1-k/m)]^{1/2}$ when crossing it. If $S_k - kS_m/m$ were replaced by its continuous time analogue, *viz.* a Brownian Bridge $B_m(t)$ for $0 \le t \le m$, the function ν on the right hand side of (6) would be replaced by 1, and this approximating expression would simplify to

$$b\varphi(b)\log[(1-t_0)t_1/t_0(1-t_1)],$$

provided in this case that $0 < t_0 < t_1 < 1$.

As indicated in Section 1, (6), which involves a one dimensional indexing set, can be proved by a much larger class of methods than are available for studying the distribution of (2). More closely related to the results of this paper is the approximation obtained independently by Siegmund (1988) and Yao (1989) to the significance level of the likelihood ratio test of H_0 against the alternative that $\mu_1 = \mu_3 \ne \mu_2$. The test statistic is

$$\max_{1 \le i < j < m} \{[S_j - S_i - (j-i)S_m/m]^2/[(j-i)(1-(j-i)/m]\}.$$

In this case, if $b \to \infty$, $m \to \infty$, while $b/m^{1/2} \to b_0$, then for any $0 \le \delta < \frac{1}{2}$ when H_0 holds,

$$P\Big\{\max_{\substack{0<i<j<m \\ \delta m < j-i < (1-\delta)m}} [|S_j - S_i - (j-i)S_m/m|/\{(j-i)[1-(j-i)/m]\}^{1/2}] > b\Big\} \sim \frac{1}{2}b^3\varphi(b)\int_{1/2}^{1-\delta} \nu^2\Big\{\frac{b_0}{[t(1-t)]^{1/2}}\Big\}\frac{dt}{t^2(1-t)^2}. \tag{7}$$

In order to state the principle results of this paper, the following additional notation is helpful. For $0 < s < t < 1$, $0 \leq \omega < 2\pi$ let

(8)
$$\xi_1 = \xi_1(s,t,\omega) = \frac{1}{2}\cos^2\omega/[s(1-s/t)],$$
$$\xi_2 = \frac{1}{2}\left\{\frac{\cos\omega}{[t(1-t)]^{1/2}} - \sin\omega\left[\frac{s}{t(t-s)}\right]^{1/2}\right\}^2.$$

(9) **Theorem.** *Suppose $b \to \infty$, $m \to \infty$, and $b/m^{1/2} \to b_0 > 0$. For any $0 \leq \delta_i < \frac{1}{4}(i = 1,2,3)$, where H_0 is true,*

$$P\left\{\max_{\substack{\delta_1 m < i < j < (1-\delta_2)m \\ j-i > \delta_3 m}} Q_{ij}^{1/2} > b\right\}$$
$$\sim (2\pi)^{-1} b^4 \exp(-\frac{1}{2}b^2) \iiint_{\substack{\delta_1 < s < t < 1-\delta_2 \\ t-s > \delta_3 \\ 0 < \omega < 2\pi}} \xi_1 \xi_2 \nu[b_0(2\xi_1)^{1/2}]\nu[b_0(2\xi_2)^{1/2}]d\omega ds dt.$$

The following result is appropriate when $\sigma^2 = \text{var}(x_n)$ is unknown and must be estimated from the data. Its proof will not be discussed in this paper. See James, James, and Siegmund (1988) for an analogous result in the simpler case of at most one change.

(10) **Theorem.** *Assume the conditions of (9) and that $0 < b_0 < 1$. Then when H_0 is true*

$$P\left\{\max_{\substack{\delta_1 m < i < j < (1-\delta_2)m \\ j-i > \delta_3 m}} Q_{i,j} > b^2 m^{-1} \sum_1^m (x_k - \bar{x}_{0,m})^2\right\}$$
$$\sim (2\pi)^{-1} b^4 (1-b_0^2)^{(m-5)/2} \iiint_{\substack{\delta_1 < s < t < 1-\delta_2 \\ t-s > \delta_3 \\ 0 < \omega < 2\pi}} \xi_1 \xi_2 \nu[b_0\{2\xi_1/(1-b_0^2)\}^{1/2}]$$
$$\nu[b_0\{2\xi_2/(1-b_0^2)\}^{1/2}]d\omega ds dt.$$

Remark. Numerical evaluation of the integrals in (9) and (10) is somewhat onerous, but becomes considerably easier if one uses the approximation (5) for the function ν defined in (4). In the simpler cases already studied this approximation is satisfactory provided m is not too small, say $m \geq 25$; and hence one expects it will also prove useful in the present case. For example, if in (7) $b = 3.5$, $m = 40$, and $\delta = 0.1$, the right hand side equals 0.0504. If the approximation (5) for ν is used the result is 0.0494. On the other hand, if ν is replaced by 1, which would be appropriate if $S_n - nS_m/m$ were in fact continuous Brownian Bridge, $B_m(t)$ for $0 \leq t \leq m$, the integral in (7) can be evaluated explicitly but the approximation is a very poor 0.25.

3. Proof of (9).

In order to prove (9) it is helpful to rewrite the definition (1) in a way which clearly exhibits each $Q_{i,j}$ as a random variable whose distribution under H_0 is χ^2 with two degrees of freedom. Let $S_n^{(m)} = S_n - nS_m/m$. Note that under H_0 $S_n^{(m)}$ $(n = 0, 1, \ldots, m)$ is a discrete time Brownian Bridge process, with mean zero and covariance function $n_1(1 - n_2/m)$ for $n_1 \leq n_2$. It is easily verified that

$$Q_{i,j} = Z_1^2 + Z_2^2,$$

where

$$Z_1 = (S_i - iS_j/j)/[i(1 - i/j)]^{1/2} = (S_i^{(m)} - iS_j^{(m)}/j)/[i(1 - i(j))]^{1/2}$$

and

$$Z_2 = S_j^{(m)}/[j(1 - j/m)]^{1/2}$$

are independent standard normal random variables under H_0. Although the dependence of Z_1 and Z_2 on i and j has been suppressed, no confusion should result since Z_1 and Z_2 will only appear in the company of other variables which indicate the appropriate indices.

From $Q_{i,j}^{1/2} = \sup_{0 \leq \omega < 2\pi} [Z_1 \cos \omega + Z_2 \sin \omega]$ it follows that $Q^{1/2}$ defined in (2) is the maximum of a three dimensional Gaussian random field. Kim and Siegmund (1989) have used such a representation in a similar problem, and it would undoubtedly work here. However, a somewhat different approach seems more natural and simpler computationally. This approach would presumably also work for Kim and Siegmund.

Let $D = \{(i_0, j_0) : \delta_1 m < i_0 < j_0 < (1 - \delta_2)m,\ j_0 - i_0 > \delta_3 m\}$, and for each $(i_0, j_0) \in D$ let $J_0 = J_0(i_0, j_0) = \{(i, j) : (i, j) \in D,\ i > i_0 \text{ and } j = j_0, \text{ or } j > j_0\}$. If $Q_{i,j} > b$ for some $(i, j) \in D$, there must be a "last" value (i_0, j_0), where "last" means that $Q_{i,j} < b$ for all $i > i_0$ and $j = j_0$ and for all $j > j_0$. Hence

$$(11)\quad P\Big\{\max_{(i,j)\in D} Q_{i,j}^{1/2} > b\Big\}$$
$$= \sum_{(i_0,j_0)\in D} \int_0^\infty \int_0^{2\pi} P\left\{Q_{i_0,j_0}^{1/2} \in b + m^{-1/2}dx, \arg(Z_1, Z_2) \in d\omega\right\}$$
$$\times P\left\{\max_{(i,j)\in J_0} Q_{i,j}^{1/2} \leq b | Q_{i_0,j_0}^{1/2} = b + m^{-1/2}x, \arg(Z_1, Z_2) = \omega\right\}.$$

Obviously

$$(12)\quad P\{Q_{i_0,j_0}^{1/2} \in b + m^{-1/2}dx, \arg(Z_1, Z_2) \in d\omega\}$$
$$= (b + m^{-1/2}x) \exp[-\frac{1}{2}(b + m^{-1/2}x)^2]m^{-1/2}dx(2\pi)^{-1}d\omega$$
$$\sim b_0 \exp[-\frac{1}{2}b^2 - b_0 x]dx(2\pi)^{-1}d\omega$$

unfiromaly for bounded x. Some technical anlaysis along the lines given in Siegmund (1988) shows that it is permissible to ignore terms (i_0, j_0) in (11) near the boundaries of D, and in analyzing the conditional probability to consider only values of (i,j) within distance $0(m^{1/8})$ of (i_0, j_0) and arbitrary fixed values of x.

The conditional probability in (11) equals

$$P\Big\{ \max_{(i,j)\in J_0} m^{1/2}(Q_{i,j}^{1/2} - Q_{i_0,j_0}^{1/2}) \le -x | Q_{i_0,j_0}^{1/2} = b + m^{-1/2}x,\ Z_1 = (b + m^{-1/2}x)\cos\omega \Big\}. \tag{13}$$

Let $f(y_1, y_2, s, t) = \{(y_1 - sy_2/t)^2/s(1-s/t) + y_2^2/t(1-t)\}^{1/2}$, so

$$Q_{i,j}^{1/2} = m^{1/2} f[m^{-1}S_i^{(m)},\ m^{-1}S_j^{(m)},\ m^{-1}i,\ m^{-1}j].$$

Moreover, a Taylor series expansion shows that if $k = i - i_0$ and $\ell = j - j_0$ are $O(m^{1/8})$, then

$$m^{1/2}(Q_{i,j}^{1/2} - Q_{i_0,j_0}^{1/2}) = (S_{i_0+k}^{(m)} - S_{i_0}^{(m)})\frac{\partial f}{\partial y_1} + (S_{j_0+\ell}^{(m)} - S_{j_0}^{(m)})\frac{\partial f}{\partial y_2} + k\frac{\partial f}{\partial s} + \ell\frac{\partial f}{\partial t} + O_p(m^{-3/4}), \tag{14}$$

where the partial derivatives are evaluated at $m^{-1}(S_{i_0}^{(m)}, S_{j_0}^{(m)}, i_0, j_0)$.

Suppose now that $m \to \infty$ with $m^{-1}(i_0, j_0) \to (s,t)$ for some $0 < s < t < 1$. A long messy calculation of means and covariances shows that under the conditioning indicated in (13) the right hand side of (14) converges in law to the random field

$$W_k + W'_\ell \quad (k, \ell = 0, \pm 1, \pm 2, \ldots),$$

where W_k and W'_ℓ are mutually independent random walks with $W_0 = W'_0 = 0$. The increments of W_k are $N(-b_0\xi_1, 2\xi_1)$ and the increments of W'_ℓ are $N(-b_0\xi_2, 2\xi_2)$, where ξ_1 and ξ_2 are defined in (8). It follows from the definition of J_0 that the conditional probability in (13) converges to

$$P\Big\{\max_{k\ge 1} W_k + \max_{\ell\ge 0} W'_\ell \le -x,\ \max_{k\le 0} W_k + \max_{\ell\ge 1} W'_\ell \le -x\Big\}. \tag{15}$$

It is necessary to remember that in (15), as well as in (16) and (17) below, the distribution of the random walks $\{W_k\}$ and $\{W'_\ell\}$ depend upon ξ_1 and ξ_2, and hence ultimately on (s, t, ω).

Since $\max_{k\le 0} W_k \ge 0$, $\max_{k\le 0} W_k + \max_{\ell\ge 1} W'_\ell \le -x$ imples $\max_{\ell\ge 1} W'_\ell \le -x$ and hence $\max_{\ell\ge 0} W'_\ell = 0$. It follows that the probability (15) equals

$$P\Big\{\max_{k\ge 1} W_k \le -x,\ \max_{k\le 0} W_k + \max_{\ell\ge 1} W'_\ell \le -x\Big\} = P\Big\{\max_{k\ge 1} W_k \le -x\Big\} P\Big\{\max_{k\le 0} W_k + \max_{\ell\ge 1} W'_\ell \le -x\Big\}. \tag{16}$$

Substituting (12) and (16) into (11) and approximating the sum over D by a double integral, one obtains

$$(17) \qquad P\left\{\max_{(i,j)\in D} Q_{i,j}^{1/2} > b\right\} \sim (2\pi)^{-1/2} b_0 m^2 \exp\left(-\frac{1}{2}b^2\right)$$
$$\times \iiint\int_0^\infty \exp(-b_0 x) P\left\{\max_{k\geq 1} W_k \leq -x\right\}$$
$$P\left\{\max_{k\geq 0} W_k + \max_{\ell\geq 1} W'_\ell \leq -x\right\} dx dw ds dt,$$

where the outer three integrals are over the domain indicated in the statement of the theorem.

The proof is concluded by appealing to Lemma (21) in Section 4 in order to evaluate the inner integral on the right hand side of (17) in terms of the function ν defined in (4).

Remarks. (i) The preceding calculation breaks down in a neighborhood of points (s, t, w) where ξ_1 or ξ_2 vanishes, e.g. along $\omega = \pi/2$. Since the totality of such points has measure 0, it is easy to see that they cause no real difficulty, although they present one more technical obstacle to a completly rigorous exposition.

(ii) In Siegmund (1988) it was necessary to evaluate an integral similar to the inner integral in (17). In fact the same method works here. Since that renewal theoretic argument is a particularly interesting feature of the method of this paper, it is given in detail in the following section. An extension to deal with higher dimensional random fields is also given.

(iii) The structure of the present problem which makes possible an evaluation of the inner integral in (17) is that (a) locally the random field behaves like a superposition of independent random walks, and (b) the constant b_0 in the exponential function is twice the ratio of the mean to the variance for the two random walks. Although (a) is at first surprising, examination of the variety of examples considered by Hogan and Siegmund (1986), Siegmund (1988), and Aldous (1989) leads one to view it as a natural consequence of the structure of the original random field. On the other hand (b) appears as the outcome of an elaborate calculation, and after comparison with previously studied examples it still seems rather mysterious.

4. Some random walk and renewal theory

In this section random walk and renewal theory are used to evaluate the inner integral in (17) in terms of the function ν defined in (4). The method comes from Siegmund (1988). An extension appropriate for higher dimensional random fields is also given. The notation is not necessarily consistent with that of Sections 1-3.

Let $W_n (n = 1, 2, \ldots, W_0 = 0)$ be a random walk with $\mu = E(W_1) > 0$. Let $\tau_+ = \inf\{n : W_n > 0\}$.

(18) Lemma. *For all $x \geq 0$*

$$\mu^{-1} P\left\{\min_{n\geq 1} W_n > x\right\} = [E(W_{\tau_+})]^{-1} P\{S_{\tau_+} > x\}.$$

Proof. This result is well-known. See, for example, Woodroofe (1982, Theorem 2.7) or Siegmund (1985, Problem 8.13). □

(19) Lemma. *If W_1 is $N(\mu, \sigma^2)$, then*

$$\int_0^\infty \exp(-2\mu x/\sigma^2) P\left\{\min_{n\geq 1} W_n > x\right\} dx = \mu\nu(2\mu/\sigma)$$
$$= \frac{1}{2}\sigma^2 P\left\{\min_{n\geq 0} S_n = 0\right\} / E(W_{\tau_+}),$$

where ν is defined in (4).

Proof. We write P_{μ,σ^2} to denote dependence of probabilities on the parameters μ, σ^2. Let $\varsigma = \mu/\sigma$. By the change of variables $y = x\sigma^{-1}$ and Lemma (18) one obtains

$$\int_0^\infty \exp(-2\mu x/\sigma^2) P_{\mu,\sigma^2}\left\{\min_{n\geq 1} W_n > x\right\} dx$$
$$= \sigma \int_0^\infty \exp(-2\varsigma y) P_{\varsigma,1}\left\{\min_{n\geq 1} W_n > y\right\} dy$$
$$= \mu[E_{\varsigma,1}(W_{\tau_+})]^{-1} \int_0^\infty \exp(-2\varsigma y) P_{\varsigma,1}\{W_{\tau_+} > y\} dy$$
$$= \mu[2\varsigma E_{\varsigma,1}(W_{\tau_+})]^{-1}[1 - E_{\varsigma,1}\exp(-2\varsigma W_{\tau_+})]$$
$$= \mu[2\varsigma E_{\varsigma,1}(W_{\tau_+})]^{-1} P_{-\varsigma,1}\{\tau_+ = \infty\},$$

where the last equality is a consequence of Wald's likelihood ratio identity. That the last three expressions all equal $\mu\nu(2\varsigma)$ is well known (e.g., Siegmund (1985, pp. 175–176)). That the last one also equals

$$\frac{1}{2}\sigma^2 P_{\mu,\sigma^2}\left\{\min_{n\geq 0} S_n = 0\right\} / E_{\mu,\sigma^2}(W_{\tau_+})$$

is obvious. □

Now let W_n' $(n = 0, 1, \ldots)$ be a second random walk, independent of the first, with $\mu' = E(W_1') > 0$. Let $M = \min_{n\geq 1} W_n$, $M^- = \min(M, 0) = \min_{n\geq 0} W_n$, and let $M' = \min_{n\geq 1} W_n'$.

(20) Lemma. *For any* $\lambda > 0$

$$\int_0^\infty \exp(-\lambda x) P\{M > x\} P\{M^- + M' > x\} dx$$
$$= \int_0^\infty \exp(-\lambda y) P\{M' \in dy\} \int_0^y P\{M > x\} \exp[\lambda(y-x)] P\{M^- > -(y-x)\} dx.$$

Proof. The result follows at once by writing

$$P\{M^- + M' > x\} = \int_x^\infty P\{M' \in dy\} P\{M^- > x - y\}$$

and inverting the order of integration.

(21) Lemma. *Suppose* W_1 *is* $N(\mu, \sigma^2)$, W_1' *is* $N(\mu', (\sigma')^2)$, *and that* $\alpha = \mu/\sigma^2 = \mu'/(\sigma')^2$. *Then*

$$\int_0^\infty \exp(-2\alpha x) P\{M > x\} P\{M^- + M' > x\} dx = 2\alpha\mu\mu' \nu(2\mu/\sigma)\nu(2\mu'/\sigma').$$

Proof. It suffices to evaluate the right hand side of (20) for $\lambda = 2\alpha$. By (18) the inner integral on the right hand side of (20) (with $\lambda = 2\alpha$) is

$$\mu \int_0^y \frac{P\{W_{\tau_+} > x\}}{E(W_{\tau_+})} \exp\{2\alpha(y-x)\} P\{M^- > -(y-x)\} dx \tag{22}$$
$$= \mu \int_0^y F_0(dx) Z(y-x),$$

where $F_0(dx) = P\{W_{\tau_+} > x\} dx / EW_{\tau_+}$ is the stationary measure for the renewal process determined by $F(dx) = P\{W_{\tau_+} \in dx\}$ and, as will be shown in Lemma (23) below,

$$Z(x) = \exp(2\alpha x) P\{M^- > -x\}$$

is a solution of the renewal equation

$$Z(y) = z(y) + F * Z(y)$$

associated with $z(y) = \exp(2\alpha y) P\{M^- = 0\}$. From standard renewal theory (e.g., Feller (1972, Chapter XI)) one knows that $Z = U * z$, where $U = \sum_0^\infty F^{n*}$ is the renewal measure associated with F. Also $F_0 * U(dx) = dx/EW_{\tau_+}$ and hence

$$F_0 * Z(y) = F_0 * U * z(y) = \int_0^y ds \exp[2\alpha(y-x)] P\{M^- = 0\}/EW_{\tau_+}$$
$$= (2\alpha EW_{\tau_+})^{-1} P\{M^- = 0\}[\exp(2\alpha y) - 1].$$

Substituting this into (22), one sees from (20) that the left hand side of (21) equals

$$\mu(2\alpha EW_{\tau_+})^{-1}P\{M^- = 0\}\int_0^\infty [1 - \exp(-2\alpha y)]P\{M' \in dy\}$$

$$= \mu(EW_{\tau_+})^{-1}P\{M^- = 0\}\int_0^\infty \exp(-2\alpha y)P\{M' > y\}dy$$

$$= 2\alpha\mu\mu'\nu(2\mu/\sigma)\nu(2\mu'/\sigma'),$$

where the final equality is a consequence of (19). □

The following result has been used above in the proof of (21).

(22) **Lemma.** *Suppose W_1 is $N(\mu, \sigma^2)$ and put $\alpha = \mu/\sigma^2$. Let $F(dx) = P\{W_{\tau_+} \in dx\}$. The function $Z(x) = \exp(2\alpha x)P\{M^- > -x\}$ satisfies the renewal equation $Z = z + F * Z$ with $z(x) = \exp(2\alpha x)P\{M^- = 0\}$.*

Proof. As above we write P_{μ,σ^2} to denote dependence of probabilities on μ, σ^2. Let $G(x) = P_{\mu,\sigma^2}\{M^- > -x\} = P_{-\mu,\sigma^2}\{\max_{n\geq 0} W_n < x\}$ $(x > 0)$, so $Z(x) = \exp(2\alpha x)G(x)$. Then

$$G(x) = P_{-\mu,\sigma^2}\{\tau_+ = \infty\} + \int_{(0,x)} P_{-\mu,\sigma^2}\{\tau_+ < \infty, W_{\tau_+} \in dy\}G(x-y).$$

Multiplying by $\exp(2\alpha x)$ yields

$$Z(x) = \exp(2\alpha x)P_{\mu,\sigma^2}\{M^- = 0\}$$

$$+ \int_{(0,x)} P_{-\mu,\sigma^2}\{\tau_+ < \infty, W_{\tau_+} \in dy\} \exp(2\alpha y)Z(x-y).$$

By Wald's likelihood ratio identity

$$P_{-\mu,\sigma^2}\{\tau_+ < \infty, W_{\tau_+} \in dy\} \exp(2\alpha y) = P_{\mu,\sigma^2}\{W_{\tau_+} \in dy\},$$

which completes the proof. □

The following result indicates how (21) can be extended to deal with higher dimensional random fields exhibiting the same structure as that encountered above. Let W_n'', $n = 0, 1, \ldots$ be a third random walk, independent of $\{W_n\}$ and $\{W_n'\}$, with W_1'' having a $N(\mu'', (\sigma'')^2)$ distribution.

(24) **Proposition.** *Assume $\alpha = \mu/\sigma^2 = \mu'/(\sigma')^2 = \mu''/(\sigma'')^2$. Let $M_1, \ldots, M_5$ be independent random variables. Suppose M_1 and M_2 have the distribution of*

$\min_{n\geq 1} W_n$, M_3 *and* M_4 *have the distribution of* $\min_{n\geq 1} W'_n$, *and* M_5 *has the distribution of* $\min_{n\geq 1} W''_n$. *Let* a^- *denote* $\min(a,0)$. *Then*

$$\int_0^\infty \exp(-2\alpha x) P\{M_1 > x\} P\{M_2^- + M_3 > x,\, M_2^- + M_4^- + M_5 > x\} dx$$
$$= (2\alpha)^2 \mu\mu'\mu''\nu(2\mu/\sigma)\nu(2\mu'/\sigma')\nu(2\mu''/\sigma'').$$

Proof. Since $P\{M_2^- + M_3 > x,\, M_2^- + M_4^- + M_5 > x\}$

$$= \int_x^\infty P\{\min(M_3, M_4^- + M_5) \in dy\} P\{M_2^- > -(y-x)\},$$

by (18) the integral on the left hand side of (24) equals

$$\mu \int_0^\infty P\{\min(M_3, M_4^- + M_5) \in dy\} \exp(-2\alpha y)$$
$$\int_0^y \frac{P\{W_{\tau_+} > x\}}{EW_{\tau_+}} \exp[2\alpha(y-x)] P\{M_2^- > -(y-x)\} dx.$$

By the proof of (21) the inner integral equals $(2\alpha EW_{\tau_+})^{-1} P\{M_1^- = 0\}[\exp(2\alpha y) - 1]$ and hence the entire expression equals

$$\mu(2\alpha EW_{\tau_+})^{-1} P\{M_1^- = 0\} \int_0^\infty P\{\min(M_3, M_4^- + M_5) \in dy\}[1 - \exp(-2\alpha y)]$$
$$= (2\alpha)\mu\nu(2\mu/\sigma) \int_0^\infty \exp(-2\alpha x) P\{M_3 > x\} P\{M_4^- + M_5 > x\} dx$$

by (19) and an integration by parts. This reduces the present case to the one previously considered, so (24) follows from (21). □

References

[1] D. Aldous, *Applications of the Poisson Clumping Heuristic*, Springer- Verlag, New York-Heidelberg-Berlin (1989).

[2] P. Bickel & M. Rosenblatt, Two-dimensional random fields, in *Multivariate Analysis -III* (P. K. Krishnaiah, ed.), Academic Press, New York (1973), 3-15.

[3] W. Feller, *An Introduction to Probability Theory and Its Applications*, Vol. II, John Wiley and Sons, New York (1972).

[4] M. L. Hogan & D. Siegmund, Large deviations for the maxima of some random fields, *Adv. Appl. Math.* **7** (1986), 2-22.

[5] B. James, K. L. James & D. Siegmund, Tests for a change-point, *Biometrika* **74** (1987), 71-84.

[6] B. James, K. L. James & D. Siegmund, Conditional boundary crossing probabilities with applications to change-point problems, *Ann. Probab.* **16** (1988), 825-839.

[7] H.-J. Kim & D. Siegmund, The likelihood ratio test for a change-point in simple linear regression, *Biometrika* **76** (1989), 409-423.

[8] J. Pickands, Upcrossing probabilities for stationary Gaussian processes, *Trans. Amer. Math. Soc.* **145** (1969), 51-73.

[9] C. Qualls & H. Watanabe, Asymptotic properties of Gaussian random fields, *Trans. Amer. Math. Soc.* **177** (1973), 155-171.

[10] D. Siegmund, *Sequential Analysis : Tests and Confidence Intervals*, Springer-Verlag, New York-Heidelberg-Berlin (1985).

[11] D. Siegmund, Approximate tail probabilities for the maxima of some random fields, *Ann. Probab.* **16** (1988), 487-501.

[12] M. Woodroofe, Frequentist properties of Bayesian sequential tests, *Biometrika* **63** (1976), 101-10.

[13] M. Woodroofe, Large deviations of the likelihood ratio statistic with applications to sequential testing, *Ann. Statist.* **6** (1978), 72-84.

[14] Qi-wei Yao, Large deviations for the boundary crossings of some random fields, *J. Math. Res. Exposition* **9** (1989), 181-192.

A way of using auxiliary randomization

Charles Stein

Abstract. For the most part this is a shorter and more systematic version of the main theoretical ideas of my 1986 book. A problem concerning random subsets of a finite group is discussed as an illustration.

1. Introduction

The use of auxiliary randomization in the study of probabilistic problems goes back at least to the reflection principle for random walk. My main aim in this paper is to expound and illustrate an approach to finite probability theory based on this simple and natural idea, which is treated in greater detail, but somewhat chaotically, in Stein (1986). Starting with Barbour (1982) a related approach has been applied extensively to combinatorial problems, in particular to the study of random graphs. His work is based on the paper by Chen (1975) on Poisson approximation, which is related to my original paper, Stein (1972), on normal approximation. Long runs and matching sequences have been studied in a similar way by Arratia, Goldstein, and Gordon (1989). Other possible applications have been suggested by Aldous in his imaginative book, Aldous (1989) and in unpublished work.

One aspect of this approach is the use of auxiliary randomization to construct an exchangeable pair of random variables. The direct use of such an exchangeable pair for the approximate computation of ratios of probabilities of different values of the random variable is described in Section 2. Some advantages and disadvantages of this direct approach are discussed at the end of that section. An application to the counting of Latin rectangles is given in Section 3.

In Section 4 I introduce a simple algebraic formalism based on such an exchangeable pair. It is easy to see that the expectation of an antisymmetric function of an exchangeable pair is zero. Thus, if TF is the conditional expectation of the antisymmetric function F of an exchangeable pair (X, X') given X, then the image of T is contained in the kernel of E, the expectation mapping for functions of X. In fact, under a simple connectedness condition, $\operatorname{im} T = \ker E$. This is related to some trivial aspects of homology theory.

In Section 5, the notion of an approximation (E_0, T_0) to the pair (E, T) is introduced. Only a very rough sketch of this idea can be given here. Typically, E_0 is a conjectured approximation to E, for example standard normal or Poisson with parameter λ. Often T_0 is a fairly obvious approximation to T and of course we have $\ker E_0 = \operatorname{im} T_0$, apart from possible technical complications. For example, when E_0

Probability Theory

is standard normal, T_0 is often given by

$$(1) \qquad (T_0 f)(w) = f'(w) - wf(w),$$

where f' denotes the derivative of f. When E_0 is Poisson with parameter λ, T_0 is usually given by

$$(2) \qquad (T_0 f)(w) = \lambda f(w+1) - wf(w).$$

By introducing some additional structure to the approximating pair (E_0, T_0), essentially solving the equation

$$(3) \qquad T_0 f = h - E_0 h$$

for f, we obtain an identity

$$(4) \qquad Eh(W) = E_0 h + E(Rh),$$

where Rh is a random variable depending linearly on h in an explicit, though perhaps complicated, way. If the choices have been successful, it will often be obvious that $E(Rf)$ is small. If we need a better approximation, we may apply a similar method to the study of $E(Rf)$ or try to treat (4) as a functional equation, or perhaps try some combination of these two approaches, or perhaps start over with a better approximation (E_0, T_0), or use a different exchangeable pair.

In Section 6 I study a different approach to the determination of $\ker E$ in the special case of a sum of indicator random variables, typically dependent. This is a slight modification of the approach of Barbour (1982). It involves the use of coupling, in which we study the small difference between the distributions of two random variables by a careful choice of the way we construct them on a common probability space. The resulting identity can be used as an alternative starting point for the formalism of Section 5. In Section 7 I apply the approach of Section 6 to a problem concerning random subsets of groups that was suggested by Persi Diaconis. Let G be a finite group with n elements and let $\{D_1 \cdots D_k\}$ be a random subset of G with k/n small. For given $x \in G$ it is proved that the number of $i, j, \in \{1 \cdots k\}$ for which $D_i D_j^{-1} = k$ has approximately a Poisson distribution. An approximation with error $O\left(\frac{k^2}{n^2} + \frac{1}{n}\right)$ is also obtained. Section 8 contains some closing remarks.

2. The use of auxiliary randomization to produce an exchangeable pair

One way to study the distribution of a random variable X that has only countably many possible values is to use auxiliary randomization to construct another random variable X' such that (X, X') is an exchangeable pair. This means that for all x_1 and x_2

$$(1) \qquad P\{X = x_1 \,\&\, X' = x_2\} = P\{X = x_2 \,\&\, X' = x_1\}.$$

If x_1 and x_2 are such that these quantities are positive, we can express this identity in terms of conditional probabilities

$$(2) \quad P\{X = x_1\}\, P\{X' = x_2 \mid X = x_1\} = P\{X = x_2\}\, P\{X' = x_1 \mid X = x_2\}$$

and divide by $P\{X = x_2\}\, P\{X' = x_2 \mid X = x_1\}$ to obtain

$$(3) \qquad \frac{P\{X = x_1\}}{P\{X = x_2\}} = \frac{P\{X' = x_1 \mid X = x_2\}}{P\{X' = x_2 \mid X = x_1\}}.$$

Suppose the exchangeable pair (X, X') connects the set Ω_1 of all x for which $P\{X = x\} > 0$ in the sense that, for any $x^{(1)}, x^{(2)} \in \Omega_1$ there exists a sequence $x_1, \ldots, x_n \in \Omega_1$ such that $x_1 = x^{(1)}$, $x_n = x^{(2)}$ and, for all $i \in \{1 \cdots n-1\}$, $P\{X = x_i \;\&\; X' = x_{i+1}\} > 0$. Then the ratio of probabilities of any two possible values for X can be computed in terms of conditional probabilities by multiplying a number of cases of (3). Since the sum of the probabilities of all possible values of X is 1, this determines all $P\{X = x\}$. Of course all this is well known, at least in the case of birth and death processes. The only possible novelty lies in the use of auxiliary randomization to construct the second random variable X' of the exchangeable pair (X, X').

The notion of a (finite) probability space with an involution is often useful in clarifying the construction of an exchangeable pair. Let Ω be a finite set, P a probability measure on the algebra of all subsets of Ω and $\gamma : \Omega \to \Omega$ an involution of the probability space (Ω, P) in the sense that

$$(4) \qquad \gamma^2 = I_\Omega,$$

the identity mapping of Ω, and

$$(5) \qquad P \circ \gamma^{-1} = P.$$

If X is a mapping on Ω to a finite set Ω_1, and $X' = X \circ \gamma$, then (X, X') is an exchangeable pair. In the first paragraph Ω_1 was taken to be such that $P\{X = w_1\} > 0$ for all $w_1 \in \Omega_1$. We are ordinarily given a finite probability space (Ω_1, P_1) and must construct Ω, P, γ, and X in such a way that $P_1 = P \circ X^{-1}$.

Let us look at two examples: a random permutation and a sum of real random variables.

Example 1.

If Π is a random permutation of $\{1 \cdots n\}$ (i.e., a random $1-1$ mapping $\Pi : \{1 \cdots n\} \to \{1 \cdots n\}$, uniformly distributed over the set S of all $n!$ such mappings) and (I, J) a random transposition of two elements of $\{1 \cdots n\}$, uniformly distributed over the set of all $\binom{n}{2}$ such transpositions, independent of Π, then the pair (Π, Π'), with $\Pi' = \Pi \circ (I, J)$ is exchangeable, and thus, for any $f : S \to \mathbb{R}$, $\big(f(\Pi), f(\Pi')\big)$ is exchangeable.

In preparation for Section 3, I shall formulate a slight generalization of this example in terms of a probability space with an involution. Let $\mathcal{P}$ be the space of all permutations of $\{1 \cdots n\}$ and let $\Omega = \mathcal{P}^2$, with a typical element denoted by (π, α). Also define Π, $A : \Omega \to \mathcal{P}$ by $\Pi(\pi, \alpha) = \pi$ and $A(\pi, \alpha) = \alpha$. Let P be any probability measure in Ω under which Π and A are independent, Π is uniformly distributed over $\mathcal{P}$, and, for all $\alpha \in \mathcal{P}$, $P\{A = \alpha\} = P\{A = \alpha^{-1}\}$. Then $\gamma : \Omega \to \Omega$, defined by $\gamma(\pi, \alpha) = (\pi \circ \alpha^{-1}, \alpha^{-1})$ is an involution of the probability space (Ω, P). Consequently (Π, Π'), with $\Pi' = \Pi \circ A^{-1}$, is an exchangeable pair, In Section 3 I shall need the case where A has the form (I, J, K), uniformly distributed over the set of all cyclic permutations of three elements of $\{1 \cdots n\}$.

Example 2.

Let $\mathcal{T}$ be a finite set and X a random function on $\mathcal{T}$ to $\mathbb{R}$ and let T be uniformly distributed over $\mathcal{T}$ independent of X. Also let X' be another random function on $\mathcal{T}$ such that

$$X'(t) = X(t) \text{ for all } t \neq T \tag{6}$$

for every $t \in \mathcal{T}$, and the pair $\big(X(T), X'(T)\big)$ is conditionally exchangeable given that $T = t$ and given also the restriction of X to the complement of $\{t\}$. Then (X, X') is exchangeable and thus also (W, W') where

$$W = \sum_{t \in \mathcal{T}} X(t) \text{ and } W' = \sum_{t \in \mathcal{T}} X'(t). \tag{7}$$

The direct use of this method without any additional theoretical apparatus has some advantages. It is easy to explain and it may permit us to use whatever approximation is appropriate rather than to force the problem into a normal or Poisson mold because the machinery is readily available for these cases. It yields pointwise approximations directly, and these are sometimes difficult to obtain by other methods. In principle it is applicable to multi-dimensional problems as easily as to one-dimensional problems.

However this approach also has some disadvantages. The use of ad hoc methods for bounding products of ratios of probabilities and for determining the normalizing constants can be tedious and inelegant. It may not be easy to turn an intuitive feeling for the conditional probabilities on the right hand side of (3) into reasonably sharp bounds for the ratios. Also, even perfect knowledge of the point-wise probabilities may not readily enable us to evaluate a probability or expectation we are interested in. For example, consider the distribution of a sum of a large number of independently identically distributed random variables having only three possible values $\{0, \alpha, \beta\}$ where α and β are incommensurable. The pointwise distribution of the sum is a disguised verson of the trinomial distribution, but this does not readily give us sharp bounds for the departure of the distribution of this sum from normality. A similar difficulty arises for the problem of obtaining good approximations to the cumulative distribution function of a χ^2 statistic for goodness of fit.

A final objection to this direct method is that when the accuracy obtained is unsatisfactory, we have no systematic guide to how to improve it. For all of these reasons, it is appropriate to search for additional theoretical structure although, in particular cases, this simple approach may be as good as a more elaborate method, or even better.

3. An application to the counting of Latin rectangles

The method of exchangeable pairs was applied directly to the problem of determining an approximate formula for the number of l by n Latin rectangles in Stein (1978). Unfortunately that paper is difficult to read and the same is true of the treatment of this problem by a slightly different method in Lectures VII and XI of Stein (1986). The present result is not as good but I hope the argument is easier to follow. Some of the first results on this problem were obtained by Erdős and Kaplansky (1946) and further results by Yamamoto (1951), both papers using the inclusion-exclusion method. The best results to date are those of Godsil and McKay (1984), who used still another approach.

An l by n Latin rectangle is an l by n array in which each of the l rows is a rearrangement of the numbers $1, \cdots, n$ and no column contains a number more than once. We shall see that the number $\nu(l,n)$ of such arrays satisfies

$$\nu(l,n) = (n!)^l e^{-\frac{l(l-1)}{2} + O\left(\frac{l^3}{n}\right)}. \tag{1}$$

This will follow easily by induction on l from (2) below. Let $L(l,n)$ by an l by n Latin rectangle and let Π be a random permutation of $\{1 \cdots n\}$, uniformly distributed over the set of all such permutations. Then the probability $p(L(l,n))$ that an $(l+1)$ by n Latin rectangle is obtained by writing $\Pi(1), \cdots, \Pi(n)$ as an $(l+1)$st row below $L(l,n)$ satisfies

$$\log p(L(l,n)) = -l + O\left(\frac{l^2}{n}\right) \tag{2}$$

Intuitively this is highly plausible. The probability that the new $(l+1)$st element written below a specified column coincides with one of those already in the column is $\frac{l}{n}$, and these events seem nearly independent. Thus the probability of no coincidence at all should be approximated well by the binomial probability $(1-\frac{l}{n})^n$, and consequently, to the indicated approximation by e^{-l}.

Some notation will be required for the proof of (2). For each $\beta \in \{1 \cdots n\}$ let $S(\beta)$ be the set of numbers in column β of $L(l,n)$ so that each $S(\beta)$ has l elements. Let W be the number of coincidences of Π with the columns of $L(l,n)$, i.e.,

$$W = \sum_{\beta=1}^{n} \{\Pi(\beta) \in S(\beta)\}. \tag{3}$$

Then

$$p(L(l,n)) = P\{W = 0\}. \tag{4}$$

We shall see that, provided n is large but l is not too large, W has approximately a binomial distribution with parameters n and l/n. Let (I, J, K) be uniformly distributed over the set of all cyclic permutations of three elements of $\{1 \cdots n\}$, independent of Π, and let $\Pi' = \Pi \circ (I, J, K)$. Then (Π, Π') is an exchangeable pair, and thus also (W, W') where W' is related to Π' as W is to Π.

In order to discuss the difference $W' - W$, let the events $C(I)$ (for creation of a coincidence at I in the transition from Π to Π') and $D(I)$ (for destruction of a coincidence at I) be defined by

$$C(I) = \{\Pi'(I) \in S(I) \;\&\; \Pi(I) \notin S(I)\} = \{\Pi(J) \in S(I) \;\&\; \Pi(I) \notin S(I)\} \tag{5}$$

and

$$D(I) = \{\Pi(I) \in S(I) \;\&\; \Pi'(I) \notin S(I)\} = \{\Pi(I) \in S(I) \;\&\; \Pi(J) \notin S(I)\}. \tag{6}$$

The events $C(J)$, $C(K)$, $D(J)$, and $D(K)$ are defined analogously. It will also be convenient to define

$$B(I) = [C(J) \cup C(K) \cup D(J) \cup D(K)]^c. \tag{7}$$

The conditional probabilities of $C(I)$ and $D(I)$ given Π are readily computed:

$$P^{\Pi}(C(I)) = \frac{(n-W)l}{n(n-1)} \tag{8}$$

and

$$P^{\Pi}(D(I)) = \frac{(n-l)W}{n(n-1)}. \tag{9}$$

The following bounds are also easy to obtain.

$$P^{\Pi}(C(I)C(J)) \le \frac{(n-W)l^2}{n(n-1)(n-2)} \tag{10}$$

$$P^{\Pi}(C(I)D(J)) \le \frac{(n-W)lW}{n(n-1)(n-2)} \tag{11}$$

and

$$P^{\Pi}(D(I)D(J)) \le \frac{(n-l)W(W-1)}{n(n-1)(n-2)} \tag{12}$$

Let

$$\bar{C}(I) = C(I)B(I)^c \tag{13}$$

$$\bar{D}(I) = D(I)B(I)^c \tag{14}$$

and so on. Then

$$\bar{C}(I) \cup \bar{C}(J) \cup \bar{C}(K) \subseteq \{W' - W = 1\} \subseteq C(I) \cup C(J) \cup C(K) \tag{15}$$

and consequently

$$\begin{aligned} &\frac{3(n-W)l}{n(n-1)}\left[1 - 4\frac{l+W}{n-2}\right] \le P^W\{W' - W = 1\} \\ &\le \frac{3(n-W)l}{n(n-1)}. \end{aligned} \tag{16}$$

Similarly

$$\begin{aligned} &\frac{3(n-l)W}{n(n-1)}\left[1 - 4\frac{l+W-1}{n-2}\right] \le P^W\{W' - W = -1\} \\ &\le \frac{3(n-l)W}{n(n-1)} \end{aligned} \tag{17}$$

We can now apply (16), (17), and (2.3) to conclude that, for all $w \in \mathbb{Z}^+$

$$\begin{aligned} &\frac{P\{W = w+1\}}{P\{W = w\}} = \frac{P\{W' = w+1 \mid W = w\}}{P\{W' = w \mid W = w+1\}} \quad \Box \\ &\le \frac{\frac{3(n-w)l}{n(n-1)}}{\frac{3(n-l)(w+1)}{n(n-1)}\left[1 - 4\frac{l+w}{n-2}\right]} = \frac{(n-w)l}{(n-l)(w+1)}\left(1 - 4\frac{l+w}{n-2}\right)^{-1} \end{aligned} \tag{18}$$

and similarly

$$\frac{P\{W = w+1\}}{P\{W = w\}} \ge \frac{(n-w)l}{(n-l)(w+1)}\left(1 - 4\frac{l+w}{n-2}\right). \tag{19}$$

Apart from the additional factors in the numerator of (19) and the denomiator of (18) these ratios are the same as for a binomial random variable, the number of successes in n independent trials with constant probability $\frac{l}{n}$ of success. Multiply (18) (and also (19)) from $w = 0$ to $w = w_1 - 1$ and replace w_1 by w, to obtain

$$P\{W = w\} = P\{W = 0\}\binom{n}{w}\left(\frac{l}{n-l}\right)^w\left(1 + O\left(\frac{w(l+w)}{n}\right)\right) \tag{20}$$

for $w + l < \frac{n}{8}$. For larger w we can use the crude inequality

$$P\{W \ge w\} \le \binom{n}{w}\frac{l}{n-1}\cdots\frac{l}{n-w+1} = \frac{l^w}{w!}. \tag{21}$$

Now (2) follows easily from (20) and (21).

4. Determination of the kernel of the expectation mapping

Direct application of the method of Section 2 is ordinarily inadequate if the aim is to study the expectation of a complicated random variable or even the probability of a complicated event. However the simple observation that the expectation of an antisymmetric function of an exchangeable pair is zero turns out to be quite useful. The reason for this will become clearer in Section 5.

Let (Ω, P) be a probability space with an involution $\gamma : \Omega \to \Omega$, as defined in Section 2. Also let Ω_1 be a finite set and X a mapping on Ω to Ω_1, and define $X' : \Omega \to \Omega_1$, by $X' = X \circ \gamma$ so that (X, X') is an exchangeable pair. Further, let P_1 be the probability measure induced in Ω_1 by X (or X'), i.e., $P_1 = P \circ X^{-1}$, let $\mathcal{X}_1$ be linear space of all functions $H : \Omega_1 \to \mathbb{R}$ and let $E_1 : \mathcal{X}_1 \to \mathbb{R}$ be the expectation mapping corresponding to P_1, so that

$$E_1 H = \sum_{\omega_1 \in \Omega_1} H(\omega_1) P_1\{\omega_1\}. \tag{1}$$

The kernel of E_1 is defined by

$$\ker E_1 = \{H : E_1 H = 0\}. \tag{2}$$

The exchangeable pair (X, X') can be used in the following way. Let $\mathcal{F}_1$ be the linear space of all functions $F : \Omega_1^2 \to \mathbb{R}$ that are antisymmetric in the sense that for all $\omega_1, \omega_2 \in \Omega_1$,

$$F(\omega_1, \omega_2) + F(\omega_2, \omega_1) = 0. \tag{3}$$

Define $T_1 : \mathcal{F}_1 \to \mathcal{X}_1$ by

$$(T_1 F)(\omega_1) = E\left[\, F(X, X') \mid X = \omega_1 \,\right] \tag{4}$$

and define the image of T_1 by

$$\operatorname{im} T_1 = \{ T_1 F : F \in \mathcal{F}_1 \}. \tag{5}$$

Theorem 1. *If the exchangeable pair (X, X') connects Ω_1 in the sense described below (2.3), then*

$$\ker E_1 = \operatorname{im} T_1. \tag{6}$$

Proof. I shall first prove the easier part of (6), that is

$$\operatorname{im} T_1 \subseteq \ker E_1. \tag{7}$$

The connectedness condition is not required for this part. Using first the exchangeability of (X, X') and then the antisymmetry of F,

$$EF(X,X') = \tfrac{1}{2}E\left[F(X,X') + F(X',X)\right] = \tfrac{1}{2}E\left[F(X,X') - F(X,X')\right] = 0. \tag{8}$$

Consequently,

$$E_1(T_1 F) = EE^X F(X,X') = EF(X,X') = 0, \tag{9}$$

which is equivalent to (7).

The slightly harder part of (6),

$$\operatorname{im} T_1 \supseteq \ker E_1, \tag{10}$$

can be proved by induction on the number of elements of Ω_1. If $|\Omega_1| = 1$, the result is trivial. If $E_1 H_1 = 0$ then $H_1 = 0 = T_1 0$. With $n \geq 1$ suppose for the induction step that (10) holds whenever $|\Omega_1| \leq n$ and consider a situation with $\Omega_1 = n+1$. Consider the graph with vertex set Ω_1 and edge set consisting of those two-element subsets $\{\omega, \omega'\}$ of Ω_1 for which $P\{X = \omega \;\&\; X' = \omega'\} > 0$. By hypothesis, this graph is connected. List the elements of $\Omega_1 = \{w_1, ..., w_{n+1}\}$ in such a way that the induced graph on $\{w_1, ..., w_n\}$ is connected and $\{w_n, w_{n+1}\}$ is an edge. Given $H_1 : \Omega_1 \to \mathbb{R}$ satisfying $E_1 H_1 = 0$, let

$$\bar{H}_1(\omega_j) = \begin{cases} 0 & \text{if } j \leq n-1, \\ H_1(\omega_{n+1}) & \text{if } j = n+1, \\ -H_1(\omega_{n+1})\frac{\pi_{n+1}}{\pi_n} & \text{if } j = n. \end{cases} \tag{11}$$

where

$$\pi_j = P_1\{\omega_j\} \text{ for all } j \leq n+1. \tag{12}$$

Then $E_1 \bar{H}_1 = 0$. Also let

$$p_{jk} = P\{X' = \omega_k \mid X = \omega_j\} \text{ for all } j,k \leq n+1. \tag{13}$$

Define the antisymmetric function $\bar{F}_1 : \Omega_1^2 \to \mathbb{R}$ by

$$\bar{F}_1(\omega_n, \omega_{n+1}) = \frac{H_1(\omega_{n+1})}{p_{n,n+1}}, \tag{14}$$

$$\bar{F}_1(\omega_{n+1}, \omega_n) = -\frac{H_1(\omega_{n+1})}{p_{n,n+1}}, \tag{15}$$

and

$$\bar{F}_1(\omega_j, \omega_k) = 0 \text{ if } \{\omega_j, \omega_k\} \neq \{\omega_n, \omega_{n+1}\}. \tag{16}$$

From (11) and (16) and the exchangeability condition $\pi_n p_{n,n+1} = \pi_{n+1} p_{n+1,n}$ it follows that $\bar{H}_1 = T_1 \bar{F}_1$. Also $(H_1 - \bar{H}_1)(\omega_{n+1}) = 0$ and $E_1(H_1 - \bar{H}_1) = 0$. Thus by the induction assumption, applied to the exchangeable pair obtained by conditioning the distribution to $\{w_1, ..., w_n\}$ and replacing any transition of the form $(X = w_j, X' = w_{n+1})$ by $(X = w_j, X' = w_j)$, there exists an antisymmetric function F_2 vanishing outside $\{\omega_1 \cdots \omega_n\}^2$ such that

$$H_1 - \bar{H}_1 = T_1 F_2. \tag{17}$$

Then with $F_1 = \bar{F}_1 + F_2$,

$$H_1 = T_1 F_1 \in \operatorname{im} T_1. \quad \square \tag{18}$$

As I pointed out in greater detail on pp. 29–31 of my book, this is essentially the well-known fact that the zero-th reduced homology group of a connected augmented simplicial complex is zero. See, for example, pp. 62–3 of Hilton and Wylie (1965). However the proof of Lemma 4 given in my book is incorrect in that I did not exercise appropriate care in the choice of ω_{n+1} (in the present notation).

This theorem tells us that, in order to compute the expectation of a random variable H, in $\mathcal{X}_1$, we need only search for a constant c, such that

$$H_1 - c_1 \in \operatorname{im} T_1 \tag{19}$$

and then we know that

$$c_1 = E_1 H_1. \tag{20}$$

At first sight this remark may not seem very helpful. However it often happens that, intuitively, T_1 is well approximated (as far as the random variables in a subspace $\mathcal{X}_0$ of $\mathcal{X}_1$ that interests us are concerned) by a much simpler linear mapping, T_0. This suggests that we study the pair (E_0, T_0), where E_0 is a conjectured approximation to $E_1 \circ \iota$, and $\iota : \mathcal{X}_0 \to \mathcal{X}_1$ is the inclusion mapping. Solution of an appropriate auxiliary problem will then lead to an explicit expression for the difference $E_1 \circ \iota - E_0$, formula (5.5).

This program will be carried out schematically in Section 5. Some of the details of the Poisson case will be given in Section 6 in a slightly different formulation.

5. The basic diagram

In order to carry out the program outlined at the end of the last section it will be useful to look at the diagram

$$
\text{(1)}\qquad
\begin{array}{ccccc}
\mathcal{F}_1 & \xrightarrow{T_1} & \mathcal{X}_1 & \xrightarrow{E_1} & \mathbb{R} \\
\uparrow{\scriptstyle \alpha} & & \uparrow{\scriptstyle \iota} & & \uparrow{\scriptstyle I_{\mathbb{R}}} \\
\mathcal{F}_0 & \underset{U_0}{\overset{T_0}{\rightleftarrows}} & \mathcal{X}_0 & \underset{\iota_0}{\overset{E_0}{\rightleftarrows}} & \mathbb{R}
\end{array}
$$

It will be easier to indicate the motivation for introducing this diagram after stating and proving the simple theorem that it illustrates.

Theorem 1. *Suppose the linear mappings in (1) satisfy the following conditions*

$$\text{(2)}\qquad E_1 \circ T_1 = 0$$

$$\text{(3)}\qquad I_{\mathcal{X}_0} = \iota_0 \circ E_0 + T_0 \circ U_0$$

and

$$\text{(4)}\qquad E_1 \circ \iota \circ \iota_0 = I_{\mathbb{R}}.$$

Then

$$\text{(5)}\qquad E_1 \circ \iota - I_{\mathbb{R}} \circ E_0 = E_1 \circ (\iota \circ T_0 - T_1 \circ \alpha) \circ U_0.$$

Proof. It follows from (2) that

$$\text{(6)}\qquad 0 = E_1 \circ T_1 \circ \alpha = E_1 \circ \iota \circ T_0 + E_1 \circ (T_1 \circ \alpha - \iota \circ T_0).$$

Transposing the last term to the left-hand side, applying the result after U_0 and using (3) and then (4),

$$\text{(7)}\qquad \begin{aligned} E_1 \circ (\iota \circ T_0 - T_1 \circ \alpha) \circ U_0 &= E_1 \circ \iota \circ T_0 \circ U_0 \\ = E_1 \circ \iota \circ (I_{\mathcal{X}_0} - \iota_0 \circ E_0) &= E_1 \circ \iota - I_{\mathbb{R}} \circ E_0. \end{aligned}$$

This is equivalent to (5). □

Next let us see how this diagram and the hypothesis of this theorem grow out of the situation studied in the last section. There $\mathcal{X}_1$ was the linear space of all real-valued functions on Ω_1, E_1 the expectation mapping corresponding to P_1, and T_1 a certain conditional expectation mapping, and (2) follows from the theorem of Section 4. Here $\mathcal{X}_0$ is a linear space of real-valued functions on another set Ω_0 (which is typically a subset of $\mathbb{R}$ or perhaps of some $\mathbb{R}^p$ and $\iota : \mathcal{X}_0 \to \mathcal{X}_1$ is defined by

$$\iota h = h \circ W \ \text{(often written } h(W)), \tag{8}$$

where W is a function on Ω_1 to Ω_0. We are interested in studying the distribution of W, which may be a sum of random variables. Also $\iota_0 : \mathbb{R} \to \mathcal{X}_0$ is defined to take constants into the corresponding constant functions on Ω_0, i.e.,

$$(\iota_0 c)(\omega) = c \text{ for all } c \in \mathbb{R} \text{ and } \omega \in \Omega_0. \tag{9}$$

Clearly assumption (4) holds in this case.

Let us look at the rest of the diagram (1) and at the assumption (3). The linear space $\mathcal{F}_0$ and the linear mapping $\alpha : \mathcal{F}_0 \to \mathcal{F}_1$ are chosen in an attempt to have the identity $E_1 \circ T_1 \circ \alpha = 0$ provide useful information about the distribution of W. Then $T_0 : \mathcal{F}_0 \to \mathcal{X}_0$ is chosen in such a way that $\iota \circ T_0$ is conjectured to be a good approximation to $T_1 \circ \alpha : \mathcal{F}_0 \to \mathcal{X}_1$. If the choices have been fortunate we shall have

$$\ker E_0 = \operatorname{im} T_0. \tag{10}$$

This implies the existence of a linear mapping $U_0 : \mathcal{X}_0 \to \mathcal{F}_0$, not necessarily unique, satisfying (3).

Finally let us try to understand how the conclusion (5) of Theorem 1 can be useful in approximating expectations. In slightly more usual notation, (5) asserts that, for all $h \in \mathcal{X}_0$,

$$E_1 h(W) = E_0 h + E_1 \left[(\iota \circ T_0 - T_1 \circ \alpha)(U_0 h)\right]. \tag{11}$$

The left-hand side of (11) is the quantity we want to approximate, and the first term on the right is the conjectured approximation. The difference $\iota \circ T_0 - T_1 \circ \alpha$ is conjectured to be small as a consequence of the construction of T_0.

The specialization of the lower row of Diagram (1) to the Poisson case is described in a somewhat different context near the end of the next section. The normal case is discussed in Lecture 2 of Stein (1986).

6. A different basic identity

A somewhat different approach was applied very effectively to the study of random graphs by Barbour (1982). This approach is closely related to earlier work on Poisson approximation by Chen (1975) and also to my original paper, Stein (1972) on normal approximation. The present version of Barbour's method grew out of a lecture that I gave to a class on his paper. A sketch of this idea is contained in Lecture VIII of Stein (1986), starting on p. 90 in (49)–(51).

The basic identity in Theorem 1 below refers to an arbitrary sum of finitely many indicator random variables. In particular cases this identity can be compared with identities characterizing various distributions to obtain approximation theorems. In particular, this is done for Poisson approximation in Theorem 3. In the next section, this will be applied to a problem concerning random subsets of a finite group.

Some notation will be needed. Let $\mathcal{T}$ be a finite set and X a random function on $\mathcal{T}$ taking on only the values 0 and 1, and also let

$$W = \sum_{t \in \mathcal{T}} X(t) \tag{1}$$

and

$$\lambda = EW. \tag{2}$$

Also let T be uniformly distributed over $\mathcal{T}$, independent of X and let X^* be a random function on $\mathcal{T}$ having as unconditional distribution the conditional distribution of X given that $X(T) = 1$. The random variable $X(T)$ is defined on an underlying sample space Ω by

$$(X(T))\,(\omega) = (X(\omega))\,(T(\omega)) \tag{3}$$

for all $\omega \in \Omega$. Finally, let

$$W^* = \sum_{t \in \mathcal{T}} X^*(t). \tag{4}$$

Lemma 1. *Under the conditions specified in the last paragraph, for every function* $f : \mathbb{Z}^+ \to \mathbb{R}$,

$$EWf(W) = \lambda E f(W^*). \tag{5}$$

Proof.

$$EWf(W) \overset{1}{=} E\left(|\mathcal{T}|E^X X(T)\right) f(W) \tag{6}$$
$$\overset{2}{=} |\mathcal{T}|EE^X X(T)f(W)$$
$$\overset{3}{=} |\mathcal{T}|EX(T)f(W)$$
$$\overset{4}{=} |\mathcal{T}|EE^{X(T)}X(T)f(W)$$
$$\overset{5}{=} |\mathcal{T}|EX(T)E^{X(T)}f(W)$$
$$\overset{6}{=} |\mathcal{T}|EX(T)E[f(W) \mid X(T)=1]$$
$$\overset{7}{=} (|\mathcal{T}|EX(T))\, E[f(W) \mid X(T)=1]$$
$$\overset{8}{=} \lambda Ef(W^*).$$

Each step is justified by a definition or by basic properties of expectation and conditional expectation. Step 1 uses the fact that, since X is independent of T and T is uniformly distributed in $\mathcal{T}$,

$$E^X X(T) = \frac{1}{|\mathcal{T}|}\sum X(t) = \frac{W}{|\mathcal{T}|}. \tag{7}$$

At step 2 I have factored out $|\mathcal{T}|$ and used the fact that

$$(E^X Y)f(W) = E^X Yf(W) \tag{8}$$

since X determines W. Step 3 uses $EE^X = E$. Step 4 uses $E = EE^{X(T)}$. Step 5 uses $E^{X(T)}X(T)Y = X(T)E^{X(T)}Y$ with $Y = f(W)$. Step 6 uses the fact that the quantities under the expectation are equal in the two possible cases $X(T) = 0$ and $X(T) = 1$. At step 7 I have factored out the constant $E[f(W) \mid X(T) = 1]$. Finally, at step 8 I have used the definition of W^* and X^* and the fact that

$$\lambda = EW = |\mathcal{T}|EX(T). \quad \square \tag{9}$$

The following trivial lemma gives some guidance for the construction of X^*, although in each special case we have to decide on how to satisfy (11) in such a way that we can effectively approximate the conditional distribution of W^* given W. It is this conditional distribution that we ordinarily use in order to obtain useful information from (5).

Lemma 2. *Let $\tilde{T}$ be a random variable taking values in $\mathcal{T}$ distributed according to*

$$P\{\tilde{T} = t\} = P\{T = t \mid X(T) = 1\} \tag{10}$$

and let $\tilde{X}$ be a random function on $\mathcal{T}$ to $\{0,1\}$, conditionally distributed given $\tilde{T}$ according to

$$(11)\qquad P\{\tilde{X} = x \mid \tilde{T} = t\} = P\{X = x \mid T = t \;\&\; X(t) = 1\}$$

for all $t \in \mathcal{T}$ and $x : \mathcal{T} \to \{0,1\}$. Then $\tilde{X}$ has the distribution specified for X^* above Lemma 1, and consequently $\tilde{W}$, defined by

$$(12)\qquad \tilde{W} = \sum_{t \in \mathcal{T}} \tilde{X}(t),$$

has the distribution specified for W^*.

Proof. For all $x : \mathcal{T} \to \{0,1\}$,

$$(13)\qquad \begin{aligned} P\{\tilde{X} = x\} &= \sum_{t \in \mathcal{T}} P\{\tilde{T} = t\} P\{\tilde{X} = x \mid \tilde{T} = t\} \\ &= P\{T = t \mid X(T) = 1\} P\{X = x \mid T = t \;\&\; X(T) = 1\} \\ &= P\{X = x \mid X(T) = 1\}. \quad \square \end{aligned}$$

For obvious reasons, I shall write T^*, X^*, and W^* rather than $\tilde{T}$, $\tilde{X}$, and $\tilde{W}$ when talking about a special case of this lemma. In applications of these two lemmas we must introduce an underlying probability space rich enough to carry X, $\tilde{T}$, and $\tilde{X}$. Often X is defined in terms of some more basic random object, say D, in which case we apply the lemma to D and an appropriate $\tilde{D}$ rather than X and $\tilde{X}$.

Next let us look at the use of these lemmas for deriving Poisson approximations. Here the key observation is that a Poisson random variable W_λ satisfies

$$(14)\qquad EW_\lambda f(W_\lambda) = \lambda E f(W_\lambda + 1)$$

for all $f : \mathbb{Z}^+ \to \mathbb{R}$ for which either side exists, and furthermore the Poisson distributions are characterized by this condition, even if required only for all bounded f. This suggests the following lemma which carries out the details of the lower row of Diagram (5.1) in the Poisson case. The proof will not be given here, but it is straightforward and can be found among the first eight pages of Lecture VIII in Stein (1986).

Lemma 3. *Define $\mathcal{X}_\lambda$ to be the space of all functions $h : \mathbb{Z}^+ \to \mathbb{R}$ satisfying*

$$(15)\qquad EW_\lambda^k |h(W_\lambda)| < \infty$$

for all $k \in \mathbb{Z}^+$, and define $\mathcal{P}_\lambda : \mathcal{X}_\lambda \to \mathbb{R}$ by

$$(16)\qquad \mathcal{P}_\lambda h = Eh(W_\lambda) = e^{-\lambda} \sum_{w=0}^{\infty} \frac{\lambda^w}{w!} h(w).$$

Also define linear mappings $T_\lambda, U_\lambda : \mathcal{X}_\lambda \to \mathcal{X}_\lambda$ by

$$(T_\lambda f)(w) = \lambda f(w+1) - wf(w) \tag{17}$$

and

$$(U_\lambda h)(w) = e^\lambda (w-1)! [\mathcal{P}_\lambda h\varphi_{w-1} - \mathcal{P}_\lambda h \mathcal{P}_\lambda \varphi_{w-1}] \tag{18}$$

where $\varphi_{w-1} : \mathbb{Z}^+ \to \mathbb{R}$ is defined by

$$\varphi_{w-1}(w') = \begin{cases} 1 & \text{if } w' \le w-1 \\ 0 & \text{otherwise.} \end{cases} \tag{19}$$

In particular $(U_\lambda h)(0) = 0$. Then

$$\mathcal{P}_\lambda \circ T_\lambda = 0, \tag{20}$$

and

$$T_\lambda \circ U_\lambda = I_{\mathcal{X}_\lambda} - \iota_\lambda \circ \mathcal{P}_\lambda \tag{21}$$

where $\iota_\lambda : \mathbb{R} \to \mathcal{X}_\lambda$ takes each real number into the corresponding constant function. Also the linear mapping $V_\lambda : \mathcal{X}_\lambda \to \mathcal{X}_\lambda$ defined by

$$(V_\lambda h)(w) = (U_\lambda h)(w+2) - (U_\lambda h)(w+1) \tag{22}$$

satisfies

$$|(V_\lambda \psi_A)(w)| \le \frac{1}{\lambda} \wedge 1 \tag{23}$$

for all $w \in \mathbb{Z}^+$ and $A \subseteq \mathbb{Z}^+$, where

$$\psi_A(w) = \begin{cases} 1 & \text{if } w \in A \\ 0 & \text{if } w \notin A. \end{cases} \tag{24}$$

The first and third lemmas can be put together to obtain an identity that is useful in discussing Poisson approximation.

Theorem 1. *Under the conditions formulated before Lemma 1, for all $h \in \mathcal{X}_\lambda$,*

$$Eh(W) - \mathcal{P}_\lambda h = \lambda E\left[(U_\lambda h)(W+1) - (U_\lambda h)(W^*)\right]. \tag{25}$$

Proof. Applying first (21) (and (17)) and then (5),

$$\begin{aligned} E\left[h(W) - \mathcal{P}_\lambda h\right] &= E\left(T_\lambda(U_\lambda h)\right)(W) = E\left[\lambda(U_\lambda h)(W+1) - W(U_\lambda h)(W)\right] \\ &= \lambda E\left[(U_\lambda h)(W+1) - (U_\lambda h)(W^*)\right]. \quad \square \end{aligned} \tag{26}$$

Corollary 1. *Under the conditions formulated before Lemma 1, for all* $A \subseteq \mathbb{Z}^+$

$$\left|P\{W \in A\} - e^{-\lambda} \sum_{w \in A} \frac{\lambda^w}{w!}\right| \leq (1 \wedge \lambda) E|W + 1 - W^*|. \tag{27}$$

Proof. Applying Theorem 1 to $h = \psi_A$ and then using (23), we obtain

$$\begin{aligned}
&\left|P\{W \in A\} - e^{-\lambda} \sum_{w \in A} \frac{\lambda^w}{w!}\right| = \left|E\psi_A(W) - \mathcal{P}_\lambda \psi_A\right| \qquad (28)\\
&=\lambda \left|E\left[(U_\lambda \psi_A)(W+1) - (U_\lambda \psi_A)(W^*)\right]\right| \leq \lambda \sup_w \left|V_\lambda \psi_A(w)\right| E\left|W + 1 - W^*\right| \\
&\leq (1 \wedge \lambda) E|W + 1 - W^*|. \quad \square
\end{aligned}$$

7. An application to random subsets of a group

An amusing problem was suggested as a possible application by Persi Diaconis. Let n and k be natural numbers and G a group with $|G| = n$. A block design with n blocks and n treatments can be obtained by choosing k distinct elements of G, say $d_1 \cdots d_k$, and creating for each $y \in G$ the block $\{d_1 y, \ldots, d_k y\}$. Then the number of blocks in which the treatments (group elements) a and b occur together is the number of pairs i, j for which $ab^{-1} = d_i d_j^{-1}$. The distribution of this number when the elements $d_1 \cdots d_k$ are chosen at random is studied here. Presumably some questions that are more closely related to practice could be studied by similar methods.

Let D be a random function on $\{1 \cdots k\}$ to G, uniformly distributed over all $(n)_k$ possibilities subject to the restriction that $D(i) \neq D(j)$ whenever $i \neq j$. For fixed $x \in G$ with $x \neq e$ (the identity element of G) let

$$W(x) = \sum_{i,j} X_{ij}, \tag{1}$$

where X_{ij} is the indicator of the event $\{D(i)[D(j)]^{-1} = x\}$. The following theorem asserts, in part, that if k/n is small and $x^2 \neq e$ then $W(x)$ has approximately a Poisson distribution with parameter λ given by

$$\lambda = EW(x) = \frac{k^2 - k}{n - 1}. \tag{2}$$

On the other hand, if $x^2 = e$ then the random matrix X is symmetric and $W/2$ has approximately a Poisson distribution with parameter $\lambda/2$.

Theorem 1. *With G, k, x, W, and λ as defined above,*

(i) *if $x^2 \neq e$ then, for all $A \subseteq \mathbb{Z}^+$,*

$$\left| P\{W(x) \in A\} - e^{-\lambda} \sum_{w \in A} \frac{\lambda^w}{w!} \right| \leq (1 \wedge \lambda) \frac{6k}{n-2}, \tag{3}$$

but

(ii) *if $x^2 = e$ then, for all $A \subseteq \mathbb{Z}^+$,*

$$\left| P\left\{ \frac{W(x)}{2} \in A \right\} - e^{-\frac{\lambda}{2}} \sum_{z \in A} \frac{(\frac{\lambda}{2})^z}{z!} \right| \leq \left(1 \wedge \frac{\lambda}{2}\right) \frac{3k}{n-2}. \tag{4}$$

Proof of (i). For brevity, I shall write W rather than $W(x)$. Let us start by constructing a random variable W^* having as its unconditional distribution the conditional distribution of W given that $X_{IJ} = 1$, where the random pair (I, J) is uniformly distributed over the set of all $k(k-1)$ ordered pairs of distinct elements of $\{1, \cdots, k\}$, independent of W. For this purpose, in the spirit of Lemma 6.1, let D^* be the random function on $\{1, \cdots, k\}$ to G defined by

$$D^*(i) = D(i) \text{ for all } i \notin \{I, J\}, \tag{5}$$

$$D^*(J) = D(J)Z, \tag{6}$$

and

$$D^*(I) = xD(J)Z, \tag{7}$$

where Z is a random element of G, conditionally uniformly distributed (given D, I, and J) over the set of all choices for which $D^*(I)$ and $D^*(J)$, defined by (6) and (7) are different from all the $D(i)$ with $i \notin \{I, J\}$. Also let

$$W^* = \sum_{i,j} X^*_{ij}, \tag{8}$$

where X^*_{ij} is the indicator of the event $\{D^*(i)[D^*(j)]^{-1} = x\}$. It is not difficult to see that W^* has the desired distribution.

It remains ony to apply Theorem 6.2 to conclude that, for any $A \subseteq \mathbb{Z}^+$,

$$\left| P\{W \in A\} - e^{-\lambda} \sum_{w \in A} \frac{\lambda^w}{w!} \right| \leq (1 \wedge \lambda) E|W + 1 - W^*|. \tag{9}$$

In order to bound $E|W+1-W^*|$, observe that each contribution of 1 to $|W+1-W^*|$ is due to an event of the type

$$X_{Ij} = 1,\ X_{Jj} = 1,\ X_{iI} = 1, \text{ or } X_{iJ} = 1 \tag{10}$$

for some i or j in $\{1\cdots k\}$, or of the type

$$X^*_{iI} = 1 \text{ or } X^*_{Jj} = 1 \tag{11}$$

for some i or j in $\{1\cdots k\}\cap\{I,J\}^c$. The probability of each of these events does not exceed $\frac{1}{n-2}$ and there are fewer than $6k$ of them. Thus

$$E|W+1-W^*| \le \frac{6k}{n-2}. \tag{12}$$

Now (3) follows from (9) and (12). □

The proof of (ii) is similar. Since the matrix X is symmetric when $x^2 = e$, it is convenient to index X by two-element subsets of $\{1\cdots k\}$ rather than by ordered pairs and write

$$X_{\{i,j\}} = X_{ij} = X_{ji}, \tag{13}$$

so that

$$W = 2\sum X_{\{i,j\}}. \tag{14}$$

Applying to $W/2$ the same argument that was applied to W in case (i), we find that

$$\left| P\left\{\frac{W}{2}\in A\right\} - e^{-\frac{\lambda}{2}}\sum_{z\in A}\frac{\left(\frac{\lambda}{2}\right)^z}{z!}\right| \le (1\wedge\lambda)E\left|\frac{W}{2}+1-\frac{W^*}{2}\right|. \tag{15}$$

Then (4) follows as in the case of (3) because there are only half as many events analogous to those in (10) and (11) for the present case. □

It may be interesting to explore the possibility of getting better results than the Poisson approximations of Theorem 1. I shall consider only case (i): $x^2 \ne e$. For this purpose we must consider the events (10) and (11) more carefully. Each of the $4k-6$ events listed in (10) contributes 1 to $W+1-W^*$ and has conditional probability $W/(k^2-k)$ given W. Each of the $2k-4$ events listed in (11) contributes -1 to $W+1-W^*$, and has conditional probability nearly equal to $\frac{1}{n}$ given W. Thus we conclude that

$$P^W\{W^* = W+2\} = \frac{2(k-2)}{n}\left(1+O\left(\frac{k+W}{n}\right)\right) \tag{16}$$

(17) $$P^W\{W^* = W+1\} = 1 - \frac{2(k-2)}{n} - \frac{(4k-6)W}{k^2-k} - O\left(\left(\frac{k+W}{n}\right)^2\right)$$

(18) $$P^W\{W^* = W\} = \frac{(4k-6)W}{k^2-k}\left(1+O\left(\frac{k+W}{n}\right)\right).$$

Then (6.22) and (6.25) yield, for all bounded $h : \mathbb{Z}^+ \to \mathbb{R}$,

(19) $$\begin{aligned} Eh(W) - \mathcal{P}_\lambda h &= \lambda E\left[(U_\lambda h)(W+1) - (U_\lambda h)(W^*)\right] \\ &= -\frac{2(k-2)}{n}\lambda E(V_\lambda h)(W) + \frac{(4k-6)}{k^2-k}\lambda EW(V_\lambda h)(W-1) + O\left(\left(\frac{k}{n}\right)^2\right). \end{aligned}$$

Applying Corollary 6.1 again, with h replaced by $V_\lambda h$ and by $w \mapsto w(V_\lambda h)(w-1)$ and using (2) and (6.23)

(20) $$Eh(W) = \mathcal{P}_\lambda h + \frac{2k}{n}\lambda \mathcal{P}_\lambda(V_\lambda h) + O\left(\frac{k^2+n}{n^2}\right).$$

It follows by a fairly tedious calculation from (VIII.36) on p. 87 of Stein (1986) that

(21) $$\mathcal{P}_\lambda(V_\lambda h) = e^{-\lambda}\sum_{w=0}^{\infty} \frac{w-(w-\lambda)^2}{2\lambda^2}\frac{\lambda^w}{w!}h(w).$$

Consequently

(22) $$Eh(W) = \mathcal{P}_\lambda h + \frac{k}{n}e^{-\lambda}\sum_{w=0}^{\infty} \frac{w-(w-\lambda)^2}{\lambda}\frac{\lambda^w}{w!}h(w) + O\left(\frac{k^2+n}{n^2}\right).$$

8. Concluding remarks

I prefer to think of this work as an outline for the solution of a broad class of problems in probability theory rather than a method of solving a more restricted class of problems, more or less mechanically, perhaps by applying some general theorems. In Section 2, we have to choose an exchangeable pair. In moderately complicated problems the appropriate choice may not be obvious. Again, in applying Section 4, it may be useful to devote quite a bit of thought to the appropriate choice of an antisymmetric function of the exchangeable pair. In applying Section 5, it may be useful to think carefully about the appropriate approximating expectation E_0 rather than automatically taking it to be normal or Poisson. The mapping $T_1 : \mathcal{F}_1 \to \mathcal{X}_1$ may suggest an approximation T_0 that is better than the one associated with the normal or Poisson approximation.

Of course this insistence on flexibility has some disadvantages. It may lead us to waste time on choices that could be made mechanically. The approach of Section 6, which is based on work of Barbour, does not require any choice analogous to that required in Sections 2 and 4. Barbour, sometimes working with Eagleson, Hall, Holst, Janson, and others, has applied this method very effectively. Lecture XIV in Stein (1986) may be relevant to the comparison of the two approaches.

There are at least three important directions for future work:

(i) It is desirable to continue working on applications, as in Barbour (1982), Arratia, Goldstein and Gordon (1989), and Stein (1978). This approach naturally leads to error bounds, sometimes fairly realistic, and often to improved approximations, as illustrated in Section 7. Of course some compromise between readability and precision of results may be necessary.

(ii) It is necessary to develop the basic theory further. To some extent this may be motivated by applications. For example, in the case of Poisson approximation for a random variable W, we are often interested mainly in the probability that $W = 0$, and, when $\lambda = EW$ is large, the error bounds of Lemma 6.3 are very poor for the purpose of evaluating this probability. Clumsy special methods were used in Stein (1978) and Stein (1986) for a similar problem. Work is also needed on multivariate normal approximation, and infinitely divisible laws, in particular stable laws.

(iii) Finally it would be interesting to try to develop a systematic course in probability theory on this basis. This would agree with the present trend away from excessive concentration on independence. As a small point, comparison of Lemma 6.1 and (14) suggests that, in some sense, the random variable W has approximately a Poisson distribution with parameter EW if and only if W^* has approximately the same distribution as $W + 1$. An attempt at such a result appears as Theorem 2 on p. 90 of Stein (1986). Barbour may have obtained similar results earlier. Analogous possibilities for normal approximation come to mind.

References

Aldous, D. (1989). *Probability Approximations via the Poisson Clumping Heuristic.* Springer-Verlag, New York.

Arratia, R., Goldstein, L. and Gordon, L. (1989). Two moments suffice for Poisson approximations: The Chen-Stein method. *Ann. Probab.* **17**, 9-25.

Baldi, P., Rinott, Y. and Stein, C. (1989). A normal approximation for the number of local maxima of a random function on a graph. In *Probability, Statistics and Mathematics : Papers in Honor of Samuel Karlin* (T. W. Anderson, K. B. Athreya and D. L. Iglehart, eds.), 59-81. Academic Press, New York.

Barbour, A. D. (1982). Poisson convergence and random graphs. *Math. Proc. Camb. Phil. Soc.* **92**, 349-359.

Barbour, A. D. and Holst, L. (1989). Some applications of the Stein-Chen method for proving Poisson convergence. *Adv. Appl. Probab.* **21**, 74-90.

Chen, L. H. Y. (1975). Poisson approximation for dependent trials. *Ann. Probab.* **3**, 534-545.

Erdős, P. and Kaplansky, I. (1946). The asymptotic number of Latin rectangles. *Amer. J. Math.* **68**, 230-236.

Godsil, C. D. and McKay, B. D. (1984). Asymptotic enumeration of Latin rectangles. *Bull. Amer. Math. Soc. (New Series)* **10**, 91-92.

Hilton, P. J. and Wylie, S. (1965). *Homology Theory : An Introduction to Algebraic Topology*. Cambridge University Press, Cambridge, U.K.

Holst, L. and Janson, S. (1989). Poisson approximation using the Stein-Chen method and coupling : Number of exceedances of Gaussian random variables. *Ann. Probab.* **18**, 713-723.

Stein, C. (1972). A bound for the error in the normal approximation to the distribution of a sum of dependent random variables. *Proc. Sixth Berkeley Symp. Math. Statist. Probab.* **2**, 583-602. Univ. California Press, Berkeley, Calif.

Stein, C. (1978). Asymptotic evaluation of the number of Latin rectangles. *J. Combin. Theory, Ser. A* **25**, 38-49.

Stein, C. (1986). *Approximate Computation of Expectations*. IMS Lecture Notes – Monograph Series Vol. 7.

Yamamoto, K. (1951). On the asymptotic number of Latin rectangles. *Japan J. Math.* **21**, 113-119.

Disintegration problems in Wiener functional integrations

Shinzo Watanabe

Introduction

Needless to say that a core of stochastic analysis is the theory of stochastic differential equations (SDE's). This method, in contrast with analytical and potential theoretic methods in probability theory based on e.g. PDE's and Dirichlet forms, provides us with trajectories of diffusion processes directly. In other words, we can obtain as a solution of SDE on a Wiener space a mapping from the Wiener space into the space of continuous trajectories of a manifold or more exactly, into the space of continuous flows of diffeomorphisms on the manifold. This mapping is often called an *Itô map* and Wiener functionals associated to it are called *Itô functionals.*

It is well known that the solution of an initial value problem of heat equations is obtained as an integral over the Wiener space of an Itô functional. On the other hand, many interesting problems in analysis, geometry and mathematical physics (e.g., index theorems, fixed point formulas, degree theorems for vector fields, Morse inequalities, etc.) are closely related to heat equations on manifolds and therefore the Wiener space integration of Itô functionals provides us with a purely probabilistic approach to the heat equation methods in these problems. Problems are concerned with the asymptotic problems of *heat kernels* rather than the solutions of initial value problems of heat equations, however, and here this probabilistic approach by Weiner space integrations meets a cruical difficulty. If we want to represent a heat kernel by an integral of an Itô functional on the Wiener space, we need to use a conditional expectation which, in a general theory of probability, is defined by the Radon–Nikodym theorem and therefore the ambiguity of an exceptional set of measure zero is inevitable. This ambiguity of exceptional set can not be allowed, however, in the above representation of heat kernels: Otherwise, we can not compute the trace of the heat kernel on the diagonal. Thus a disintegration theory, i.e. a refinement of conditional expectations, is essentially necessary in the above probabilistic approach. A method exploited so far is to use *pinned diffusion processes* (cf. Molchanov [25], Azencott [2], Hsu [8], etc.). In this approach, however, a pinned diffusion should be defined once again as a probability over continuous paths on the manifold by using the heat kernel. So there exists certainly a tautology and this approach is not quite connected with the method of SDE's. Indeed, some analytical properties of heat kernels in PDE theory are more or less required and therefore applications are usually limited to non-degenerate cases.

Probability Theory

In 1976, Malliavin ([21]) introduced a new idea of applying a functional analysis on Wiener space to Itô functionals. This theory of *Malliavin calculus* proves a better probabilistic approach to the disintegration theory for Wiener space integrations. Actually, for a finite dimensional Wiener map which is regular in the sense of Malliavin, we can define a nice induced meausre on each leaf of the finite co–dimensional foliation on the Wiener space defined by the map. It is important to notice here that every Itô map and Itô functional (more generally every Wiener functional smooth in the sense of Malliavin) can be regarded as random elements over the induced probability space on the leaf. Heat kernels are precisely represented by this disintegration and their asymptotics can be computed by using a theory of *generalized Wiener functional integrations* analogous to the Schwartz theory of distributions.

1. Sobolev spaces Wiener functionals and generalized Wiener functionals

Here we wish to formulate a framework of the Malliavin calculus as a kind of Schwartz distribution theory over Wiener space. The reader is recommended to see the exposition of K. Itô in this proceedings (also K. Itô [12]) where a very similar formulation is given in a more general setting. It should also be remarked that a much more general distribution theory has been introduced by T. Hida and others under the name of white noise analysis (cf. Meyer–Yan [23], [24]). We will discuss the case of the standard Wiener space (W,P), although the theory can be developed in the same way over any abstract Weiner space. So let $W(=W_0^r) = \{w \in C([0,\infty) \to R^r); w(0) = 0\}$ be the space of r–dimensional continuous paths starting at the origin endowed with the topology of uniform convergence over bounded intervals and P be the standard r–dimensional Wiener measure defined on the P–completion of topological σ–field on W. By a *Wiener functional*, we mean a P–measurable function defined on W. As usual, we identify two Wiener functionals with values in a same space whenever they coincide with each other P–almost surely (a.s.). Strictly speaking therefore, a Wiener functional is an equivalence class of P–measurable functions on W coinciding with each other P–a.s. The most important function spaces formed of Wiener functionals are L^p–spaces: Let E be a real separable Hilbert space and let $\mathcal{L}_p(E)$, $1 < p < \infty$, be the real Banach space formed of E–valued Wiener functionals F with finite L_p–norm $\|F\|_p = \{\int |F(w)|_E^p P(dw)\}^{1/p}$. When $E = R$, it is denoted simply by $\mathcal{L}_p$. Let $\mathcal{L}_{\infty-}(E) = \bigcap_{1<p<\infty} \mathcal{L}_p(E)$ and $\mathcal{L}_{1+}(E) = \bigcup_{1<p<\infty} \mathcal{L}_p(E)$. Then $\mathcal{L}_{\infty-}(E)$ is a Frechet space and its dual can be identified with $\mathcal{L}_{1+}(E)$ as usual. Also $\mathcal{L}_{\infty-}$ is an algebra; if $f,g \in \mathcal{L}_{\infty-}$ then $f \cdot g \in \mathcal{L}_{\infty-}$.

Let $\tau : L_2([0,\infty) \to R^r; dt) \to W$ be defined by

$$\tau(f)(t) = \int_0^t f(s)ds, \quad f \in L_2([0,\infty) \to R^r; dt)$$

and set $H = \tau\{L_2([0,\infty) \to R^r; dt)\}$ with the induced inner product

$$\langle \tau(f), \tau(g) \rangle_H = (f,g)_{L_2([0,\infty) \to R^r; dt)}.$$

Then H is a real Hilbert space densely included in W with continuous injection. H is called the *Cameron-Martin subspace* of W. For $h \in H$, $h = \tau(f)$ with $f \in L_2([0,\infty) \to R^r; dt)$, let $[h](w) \in \mathcal{L}_2$ be the *generalized inner product*, i.e. it is defined by usual stochastic integral (Wiener integral)

$$[h](w) = \int_0^\infty f(t) \cdot dw(t) = \sum_{\alpha=1}^r \int_0^\infty f^\alpha(t) dw^\alpha(t). \tag{1.1}$$

Then $C_1 = \{[h](w); h \in H\}$ is a closed subspace of $\mathcal{L}_2$ (the first order Wiener chaos) which is a mean 0 Gaussian system with covariance coinciding with the H-inner product.

From now on we follow the same route as Itô ([12]) by taking C_1 as the basic Segal space. For completness, we repeat this in our notation. A real valued Wiener functional $F(w)$ is called a *polynomial functional* if there exist a real polynomial $p(t_1, \ldots, t_n)$, and $h_1, \ldots, h_n \in H$ such that $F(w) = p([h_1](w), \ldots, [h_n](w))$. The totality of polynomial functionals is denoted by $\mathcal{P}$. More generally, for a separable real Hilbert space E, an *E-valued polynomial functional* is an E-valued Wiener functional $F(w)$ expressible in the form of a finite sum

$$F(w) = \sum F_i(w) e_i, \quad F_i \in \mathcal{P}, \quad e_i \in E.$$

The totality of E-valued polynomial functionals is denoted by $\mathcal{P}(E)$. Then $\mathcal{P}(E) \subset \mathcal{L}_p(E)$ densely for every $1 < p < \infty$.

For $F \in \mathcal{P}(E)$ and $h \in H$, the *directional derivative* $D_h F \in \mathcal{P}(E)$ *of* F *(in the direction of* h*)* is defined by

$$D_h F(w) = \lim_{\epsilon \to 0} \frac{F(w + \epsilon h) - F(w)}{\epsilon}. \tag{1.2}$$

Clearly, there exists unique $DF \in \mathcal{P}\ (H \otimes E)$ ($H \otimes E$ is the Hilbert space of linear maps $H \to E$ of Hilbert-Schmidt ($H.$-$S.$) type endowed with $H.$-$S.$ norm) such that

$$D_h F(w) = DF(w)\{h\}, \quad h \in H. \tag{1.3}$$

If $G \in \mathcal{P}\ (H \otimes E)$ and it is expessible in the form of a finite sum

$$G(w) = \sum_{i,j} G_{ij}(w) \cdot h_i \otimes e_j, \quad G_{ij} \in \mathcal{P}, h_i \in H, e_j \in E \tag{1.4}$$

where $H_i \otimes e_j \in H \otimes E$ is defined by

$$(h_i \otimes e_j)\{h\} = \langle h, h_i \rangle_H e_j, \quad h \in H,$$

we define $[G](w) \in \mathcal{P}(E)$ by

$$[G](w) = \sum_{i,j} G_{ij}(w) \cdot [h_i](w) \cdot e_j. \tag{1.5}$$

$[G]$ is well defined independently of the way of expressing it in the form (1.4). In (1.4), $\{h_i\}$ can be chosen orthonormal and then trace $DG(w) \in \mathcal{P}(E)$ is given by

$$\text{trace}\, DG(w) = \sum_{i,j} DG_{ij}(w)\{h_i\} \cdot e_j. \tag{1.6}$$

Define, for $G \in \mathcal{P}(H \otimes E)$ expressible in the form (1.4), $D^*G \in \mathcal{P}(E)$ by

$$D^*G(w) = -\text{trace}\, DG(w) + [G](w). \tag{1.7}$$

By an integration by parts in a finite dimensional Gaussian measure integration, we can prove easily

$$\int_W \langle DF(w), G(w)\rangle_{H\otimes E} P(dw) = \int_W \langle F(w), D^*G(w)\rangle_E P(dw) \tag{1.8}$$

for $F \in \mathcal{P}(E)$ and $G \in \mathcal{P}(H \otimes E)$ expressible in the form (1.4). If $F \in \mathcal{P}(E)$, then $DF \in \mathcal{P}(H \otimes E)$ is always expressible in the form (1.4). Define $LF \in \mathcal{P}(E)$ by

$$LF(w) = -D^*(DF)(w). \tag{1.9}$$

The operator $L : \mathcal{P}(E) \to \mathcal{P}(E)$ is called the *Ornstein–Uhlenbeck operator.*

Now $\mathcal{L}_2(E)$ is a Hilbert space and is decomposed into a direct sum of mutually orthogonal subspaces of Wiener's homogeneous chaos :

$$\mathcal{L}_2(E) = \mathcal{C}_0(E) \oplus \mathcal{C}_1(E) \oplus \ldots \mathcal{C}_n(E) \oplus \ldots \tag{1.10}$$

The projection operator of $\mathcal{L}_2(E)$ onto $\mathcal{C}_n(E)$ is denoted by J_n. If $F \in \mathcal{P}(E)$, then $J_nF \in \mathcal{P}(E)$ and $F = \sum J_nF$ is actually a finite sum. It holds that (cf. Itô's lecture for simple and elegant proof)

$$LF = \sum_n (-n)J_nF, \quad F \in \mathcal{P}(E). \tag{1.11}$$

Therefore, for every $s \in R$, an operator $(I-L)^s : \mathcal{P}(E) \to \mathcal{P}(E)$ is defined by

$$(I-L)^sF = \sum_{n=0}^{\infty} (1+n)^s J_nF. \tag{1.12}$$

For $1 < p < \infty$ and $s \in R$, define a norm $\|\ \|_{p,s}$ on $\mathcal{P}(E)$ (denoted also by $\|\ \|_{p,s;E}$ to make the value space clear) by

$$\|F\|_{p,s} = \|(I-L)^{s/2}F\|_p, \quad F \in \mathcal{P}(E). \tag{1.13}$$

These norms have the following basic properties (cf. e.g. [28], [35], [10]),

(a) (monotonicity)

$$\|F\|_{p,s} \leq \|F\|_{p',s'}, \text{ for any } F \in \mathcal{P}(E) \text{ if } s \leq s' \text{ and } p \leq p'.$$

(b) (compatibility) If $\{F_n\} \subset \mathcal{P}(E)$ satisfies $\|F_n - F_m\|_{p,s} \to 0$ as $n, m \to \infty$ and $\|F_n\|_{p',s'} \to 0$ as $n \to \infty$, then $\|F_n\|_{p,s} \to 0$ as $n \to \infty$.

(c) (duality) For $G \in \mathcal{P}(E)$ and $1 < p < \infty$, $s \in R$,

$$\|G\|_{p,s} = \sup\{\int_W \langle G(w), F(w)\rangle_E P(dw);\ F \in \mathcal{P}(E),\ \|F\|_{q,-s} \leq 1\} \tag{1.14}$$

where $p^{-1} + q^{-1} = 1$.

(d) (Meyer's equivalence of norms) For every $1 < p < \infty$, $s \in R$ and $k = 0, 1, 2, \ldots$, there exist positive constants $c = c_{p,s,k}$ and $C = C_{p,s,k}$ such that

(1.15)

$$c\|D^k F\|_{p,s;\underbrace{H \otimes \ldots \otimes H}_{k} \otimes E} \leq \|F\|_{p,s+k;E}$$

$$\leq C \sum_{\ell=0}^{k} \|D^\ell F\|_{p,s;\underbrace{H \otimes \ldots \otimes H}_{\ell} \otimes E}$$

for every $F \in \mathcal{P}(E)$.

Now we are in a position to define the Sobolev spaces formed of Wiener functionals and generalized Wiener functionals. Just as in Schwartz's theory of distributions, we define these objects as linear functionals on $\mathcal{P}(E)$. This idea is due to Itô ([12]). We remark that in the white noise analysis by Hida and others, $\mathcal{P}(E)$ should be replaced by a smaller space by restricting $h \in H$ to $h \in H_0$, H_0 being a properly chosen dense subspace of H. Let $\mathcal{P}'(E)$ be the real vector space formed of all linear maps $T : \mathcal{P}(E) \to R$, i.e. the algebraic dual of $\mathcal{P}(E)$. For $F \in \mathcal{P}(E)$, define $T_F \in \mathcal{P}'(E)$ by

$$T_F(G) = \int_W \langle F(w), G(w)\rangle_E P(dw), \quad G \in \mathcal{P}(E). \tag{1.16}$$

Clearly $F \to T_F$ is one to one and we can identify T_F with F. Under this identification, therefore, we have $\mathcal{P}(E) \subset \mathcal{P}'(E)$. Define for $p \in (1, \infty)$ and $s \in R$,

$$\mathcal{D}_p^s(E) = \{T \in \mathcal{P}'(E);\ \|T\|_{p,s} < \infty\} \tag{1.17}$$

where

$$\|T\|_{p,s} = sup\{T(G);\ G \in \mathcal{P}(E),\ \|G\|_{q,-s} \leq 1\},\ p^{-1} + q^{-1} = 1. \tag{1.18}$$

By the above identification of T_F and and F, $\mathcal{P}(E) \subset \mathcal{D}_p^s(E)$ and the norms defined on them are compatible; the definitions (1.18) and (1.13) coincide by the duality

(1.14). Furthermore, $\mathcal{P}(E)$ is dense in $\mathcal{D}_p^s(E)$, that is, $\mathcal{D}_p^s(E)$ coincides with $\| \quad \|_{p,s}$-closure of $\mathcal{P}(E)$. (As before, $\| \quad \|_{p,s}$ is also denoted by $\| \quad \|_{p,s;E}$).

The above properties (a) $\sim$ (d) of the norm imply the following :

$$\mathcal{D}_p^0(E) = \mathcal{L}_p(E), \tag{1.19}$$

$$\mathcal{D}_{p'}^{s'}(E) \subset \mathcal{D}_p^s(E) \text{ if } p \leq p' \text{ and } s \leq s' \text{ with continuous inclusion,} \tag{1.20}$$

$$\mathcal{D}_p^s(E)' = \mathcal{D}_q^{-s}(E),\ p^{-1} + q^{-1} = 1. \tag{1.21}$$

The operator $D : \mathcal{P}(E) \to \mathcal{P}(H \otimes E)$ can be extended uniquely to a continous operator $D : \mathcal{D}_p^{s+1}(E) \to \mathcal{D}_p^s(H \otimes E)$ for every $p \in (1, \infty)$ and $s \in R$. Its dual D^* is therefore a continuous operator $\mathcal{D}_p^{s+1}(H \otimes E) \to \mathcal{D}_p^s(E)$ for every $p \in (1, \infty)$ and $s \in R$ which is an extension of that given by (1.7). Also $L : \mathcal{P}(E) \to \mathcal{P}(E)$ can be extended uniquely to a continuous operator $L : \mathcal{D}_p^{s+2}(E) \to \mathcal{D}_p^s(E)$ for every $p \in (1, \infty)$ and $s \in R$ and (1.9) still holds for these extended operators.

We set

$$\mathcal{D}_{\infty-}^{\infty}(E) = \bigcap_{s>0} \bigcap_{1<p<\infty} \mathcal{D}_p^s(E),\ \mathcal{D}_{1+}^{-\infty}(E) = \bigcup_{s>0} \bigcup_{1<p<\infty} \mathcal{D}_p^{-s}(E). \tag{1.22}$$

$\mathcal{D}_{\infty-}^{\infty}(E)$ is a Frechet space and $\mathcal{D}_{1+}^{-\infty}(E)$ is its dual. The spaces $\mathcal{D}_p^s(E)$, $\mathcal{D}_{\infty-}^{\infty}(E)$ and $\mathcal{D}_{1+}^{-\infty}(E)$ are denoted simply by $\mathcal{D}_p^s$, $\mathcal{D}_{\infty-}^{\infty}$ and $\mathcal{D}_{1+}^{-\infty}$ respectively when $E = R$. If $\varphi : R^d \to R^{d'}$ is C^∞ with derivatives of polynomial growth and $F \in \mathcal{D}_{\infty-}^{\infty}(R^d)$, then the composite $\varphi \circ F$ belongs to $\mathcal{D}_{\infty-}^{\infty}(R^{d'})$. For chain rules for differential operators D, D^* and L, cf. [10] and [35].

Let E_1 and E_2 be real separable Hilbert spaces. For $p, q \in (1, \infty)$ and $k = 0, 1, 2, \ldots$ such that $p^{-1} + q^{-1} = r^{-1} < 1$, there exists a positive constnat $C = C_{p,q,k}$ such that

$$\|F \otimes G\|_{r,k;E_1 \otimes E_2} \leq C \|F\|_{p,k;E_1} \cdot \|G\|_{q,k;E_2} \quad \text{for every } F \in \mathcal{D}_p^k(E_1) \text{ and } G \in \mathcal{D}_q^k(E_2). \tag{1.23}$$

From this, it follows that $\mathcal{D}_{\infty-}^{\infty}$ is an algebra. More generally, if $F \in \mathcal{D}_{\infty-}^{\infty}$ and $G \in \mathcal{D}_{\infty-}^{\infty}(E)$ then $F \cdot G \in \mathcal{D}_{\infty-}^{\infty}(E)$. Thus, $\mathcal{D}_{\infty-}^{\infty}(E)$ is $\mathcal{D}_{\infty-}^{\infty}$-module. ALso, if $F \in \mathcal{D}_{\infty-}^{\infty}$ and $\Phi \in \mathcal{D}_{1+}^{-\infty}(E)$, then $F \cdot \Phi \in \mathcal{D}_{1+}^{-\infty}(E)$ is defined by

$$\langle F \cdot \Phi, G \rangle = \langle \Phi, F \cdot G \rangle \text{ for every } G \in \mathcal{D}_{\infty-}^{\infty}(E) \tag{1.24}$$

where $\langle , \rangle$ is the canonical bilinear form (natural coupling) on $\mathcal{D}_{1+}^{-\infty}(E) \times \mathcal{D}_{\infty-}^{\infty}(E)$. Thus $\mathcal{D}_{1+}^{-\infty}(E)$ is also a $\mathcal{D}_{\infty-}^{\infty}$–module. The continuity of this multiplication is more precisely stated in the following inequality which immediately follows from (1.23) : For every $p, q \in (1, \infty)$ and $k = 0, 1, 2, \ldots$ such that $p^{-1} + q^{-1} = r^{-1} < 1$, there exists a positive constant $C' = C'_{p,q,k}$ such that

$$\|F \cdot \Phi\|_{r,-k;E} \leq C' \|F\|_{p,k} \|\Phi\|_{q,-k;E} \quad \text{for every } F \in \mathcal{D}_p^k \text{ and } \Phi \in \mathcal{D}_q^{-k}(E). \tag{1.25}$$

If 1 is the Wiener functional taking the constant value 1, then clearly $1 \in \mathcal{D}^{\infty}_{\infty -}$. For $\Phi \in \mathcal{D}^{-\infty}_{1+}$, the natural coupling $\langle \Phi, 1\rangle$ is denoted by $E(\Phi)$ and is called the *generalized expectation.* $E(\Phi)$ coincides with the ordinary expectation $\int_W \Phi(w)P(dw)$ when $\Phi \in \mathcal{L}_{1+} \subset \mathcal{D}^{-\infty}_{1+}$. Note that for every $F \in \mathcal{D}^{\infty}_{-\infty}$ and $\Phi \in \mathcal{D}^{-\infty}_{1+}$, the generalized expectation $E(F \cdot \Phi)$ coincides with the natural coupling $\langle \Phi, F\rangle$. It is clear from the definition that

$$|E(\Phi)| \leq C\|\Phi\|_{p,-k}, \quad \Phi \in \mathcal{D}^{-k}_{p} \tag{1.26}$$

where the constant C is given by $C = \|1\|_{q,k}$ with $p^{-1} + q^{-1} = 1$.

2. Pull–back of Schwartz distributions by finite dimensional Wiener maps

Let $F : W \to R^d$ be a d–dimensional Wiener functional, or in a more geometric term, a d–dimensional Wiener map. It is said to be *regular* (*in the sense of Malliavin*) if it satisfies the following two conditions :

$$F \in \mathcal{D}^{\infty}_{\infty -}(R^d), \text{ i.e., } F = (F^1, \ldots, F^d) \text{ with } F^i \in \mathcal{D}^{\infty}_{\infty -}, \tag{2.1}$$

$$[\det \sigma(w)]^{-1} \in \mathcal{L}_{\infty -} \tag{2.2}$$

where the Wiener functional $\sigma(w) = (\sigma^{ij}(w))$ with values in the space of all $d \times d$ non–negative definite matrices is defined by

$$\sigma^{ij}(w) = \langle DF^i(w), DF^j(w)\rangle_H, \quad i, j = 1, \ldots, d. \tag{2.3}$$

Here, $H \otimes R$ is identified with H in the usual way. Clearly $\sigma^{ij}(w) \in \mathcal{D}^{\infty}_{\infty -}$. $\sigma(w) = \sigma_F(w)$ is called the *Malliavin covariance of F*. If F is regular, then $\sigma^{-1}(w) = (\gamma^{ij}(w))$ satisfies $\gamma^{ij}(w) \in \mathcal{D}^{\infty}_{\infty -}$.

Suppose that we are given a regular d–dimensional Wiener map F. We show that every Schwartz distribution $T(x)$ on R^d can be lifted or pulled back to a generalized Wiener functional $T \circ F$ (denoted also by $T(F)$) in $\mathcal{D}^{-\infty}_{1+}$ under the Wiener map $F : W \to R^d$. For this we introduce the following family of real Banach spaces of functions and generalized functions on R^d. Let $\mathcal{S}(R^d)$ be the real Schwartz space of rapidly decreasing C^∞–functions on R^d and define an increasing family of norms on it by

$$\|\varphi\|_{2k} = \|(1 + |x|^2 - \Delta)^k \varphi\|_\infty, \quad \varphi \in \mathcal{S}(R^d), \quad k = 0, \pm 1, \pm 2, \ldots,$$

where $\| \quad \|_\infty$ is the supremum norm and $\Delta = \sum_{i=1}^{d} (\partial/\partial x^i)^2$. Let $\mathcal{T}_{2k}$ be the closure in $\mathcal{S}'(R^d)$ of $\mathcal{S}(R^d)$ by the norm $\| \quad \|_{2k}$. Then

$$\mathcal{S}(R^d) \subset \ldots \subset \mathcal{T}_2 \subset \mathcal{T}_0 \subset \mathcal{T}_{-2} \subset \ldots \subset \mathcal{S}'(R^d)$$

with continuous inclusions and $\mathcal{T}_0 = \hat{C}(R^d)$:= the Banach space of all real continuous functions on R^d tending to 0 at infinity endowed with the supremum norm. Furthermore, $\cap_{k=1}^{\infty} \mathcal{T}_{2k} = S(R^d)$ and $\cup_{k=1}^{\infty} \mathcal{T}_{-2k} = S'(R^d)$.

Theorem 2.1. ([10],[35]) *Let $F : W \to R^d$ be a regular d-dimensional Wiener map. Then for every $p \in (1,\infty)$ and $k = 0,1,2,\ldots$, there exists a positive constant $C = C_{p,k,F}$ such that*

$$\|\varphi \circ F\|_{p,-2k} \leq C\|\varphi\|_{-2k} \quad \textit{for all } \varphi \in S(R^d). \tag{2.4}$$

Hence the map $\varphi \in S(R^d) \to \varphi \circ F \in \mathcal{D}_{\infty-}^{\infty}$ can be extended uniquely to a linear map

$$T \in S'(R^d) \to T \circ F \in \mathcal{D}_{1+}^{-\infty}$$

such that its restriction $T \in \mathcal{T}_{-2k} \to T \circ F \in \mathcal{D}_p^{-2k}$ is continuous for every $p \in (1,\infty)$ and $k = 0,1,2,\ldots$. In particular, $T \circ F \in \mathcal{D}_{\infty-}^{-\infty} := \cup_{k=1}^{\infty} \cap_{1<p<\infty} \mathcal{D}_p^{-k}$.

$T \circ F$, denoted also by $T(F)$, is called the *composition* of a *Schwartz distribution* $T \in S'(R^d)$ *and* F, or *pull-back of* T *under the Wiener map* F. Note that $\mathcal{D}_{\infty-}^{-\infty}$ is much smaller than $\mathcal{D}_{1+}^{-\infty}$ and for any G in $\mathcal{D}_{1+}^{\infty} := \cap_{k=1}^{\infty} \cup_{1<p<\infty} \mathcal{D}_p^k$, which is much larger than $\mathcal{D}_{\infty-}^{\infty}$, $G \cdot T \circ F \in \mathcal{D}_{1+}^{-\infty}$ is well defined and hence the generalized expectation $E(G \cdot T \circ F)$ is well defined.

Example. For one-dimensional Wiener space, let

$$G(w) = \exp\{\alpha \int_0^1 w(s)^2 ds\}, \quad \alpha \in R.$$

Then $G \in \mathcal{D}_{1+}^{\infty}$ if and only if $\alpha < \pi^2/8$ and if $f(x)$ is a C^∞ – function with compact support, then $G(w)f(w(1)) \in \mathcal{D}_{1+}^{\infty}$ if and only if $\alpha < \pi^2/2$. Clearly, $G \in \mathcal{D}_{\infty-}^{\infty}$ if and only if $\alpha \leq 0$.

Remark. It is Kuo [17] who first defined $\delta(w(1))$, the composite of the Dirac δ–function and $w(1)$ on one dimensional Wiener space, as a generalized Brownian functional in the framework of white noise analysis by Hida. He named it as Donsker's δ–function. It is a formal density of the measure of Brownian bridge with respect to the Wiener measure. On the other hand, Brownian bridge is obtained as an image measure, under a simple linear map on the Wiener space, of Wiener measure. Therefore, Kuo's work suggests us that it is very important to work a similar problem in the case of more general non–linear mappings, Itô maps, in particular. Our theorem above is an answer to it. We think that an efficiency of the Malliavin calculus is manifestly shown in this kind of problems.

3. The regularity of conditional expectations and conditional probability laws

Let $F : W \to R^d$ be a d-dimensional Wiener map which we assume to be regular in the sense of (2.1) and (2.2). Noting that δ_x, the Dirac δ-function at $x \in R^d$, is in $\mathcal{T}_{-2m}$ for $m \geq m_0$ where $m_0 = [d/2] + 1$, we see from Th. 2.1 that, for $k = 0, 1, \dots$ and $1 < p < \infty$

$$x \in R^d \to \delta_x(F) \in \mathcal{D}_p^{-2m_0-2k}$$

is continuously differentiable $2k$-times. Hence, for every $G \in \bigcup_{1<q<\infty} \mathcal{D}_q^{2m_0+2k}$,

$$x \in R^d \to E[G \cdot \delta_x(F)]$$

is C^{2k} and therefore it is C^∞ if $G \in \mathcal{D}_{1+}^\infty$. In particular,

$$p_F(x) = E[\delta_x(F)] \tag{3.1}$$

is a C^∞-function on R^d. But this $p_F(x)$ is the density, with respect to the Lebesgue measure on R^d, of the image measure $P \circ F^{-1}$ (the probability law of the d-dimensional Wiener functional F), as is seen from

$$\int_{R^d} p_F(x)\varphi(x)dx = \int_{R^d} \varphi(x)E[\delta_x(F)]dx = E[\int_{R^d} \varphi(x)\delta_x(F)dx] = E[\varphi(F)].$$

In this way, we have deduced that the law of F has a C^∞-density if F is regular. Furthermore, it is easy to see that $E[G \cdot \delta_x(F)]$ is a version of $E[G \mid F = x]p_F(x)$. Therefore, by defining the *strict support* $SSP(F)$ *of* F by

$$SSP(F) = \{x \in R^d;\ p_F(x) > 0\}, \tag{3.2}$$

we see that the conditional expectation of $G \in \mathcal{D}_{1+}^\infty$ given F has a smooth version on $SSP(F)$.

We can proceed one step further to show that the conditional probability is precisely defined on each leaf $\{w; F(w) = x\}$, $x \in SSP(F)$, of the foliation on W defined by F. For this, we need to introduce the notion of *capacities* and *regularization of Wiener functionals* due to Malliavin [22] (cf. [29]). *In the following, we always assume that* $p \in (1, \infty)$ *and* $r > 0$. For an open set $0 \subset W$, we define its *(p,r)-capacity* $C_{p,r}(0)$ by

$$C_{p,r}(O) = \inf\{\|F\|_{p,r};\ F \in \mathcal{D}_p^r,\ F(w) \geq 1 \text{ a.e. on } O\} \tag{3.3}$$

and for any $A \subset W$, its *(outer) (p,r)-capacity* $C_{p,r}(A)$ by

$$C_{p,r}(A) = \inf\{C_{pr,}(O);\ A \subset O, O; \text{open}\} \tag{3.4}$$

A function F is defined on the Wiener space with values in a topological space is called *(p,r)–quasi continuous* if for every $\epsilon > 0$ there exists an open set $O \subset W$ with $C_{p,r}(O) < \epsilon$ and $F|_{W\setminus O}$ is continuous. A Wiener functional F with values in a topological space is called to *possess (p,r)–quasi continuous modification* $\tilde{F}$ if among the equivalence class of P–measurable functions defining F, we can choose a (p,r)–quasi continuous function $\tilde{F}$. If $\tilde{F}$ and $\tilde{F}'$ are two (p,r)–quasi continuous modifications of F, then they coincide everywhere except on a set of (p,r)–capacity 0. A set $A \subset W$ is called *slim* if $C_{p,r}(A) = 0$ for every $1 < p < \infty$ and $r > 0$ and a function on W with values in a topological space is called *∞–quasi continuous* if it is (p,r)–quasi continuous for every $1 < p < \infty$ and $r > 0$. The notion of *∞–quasi continuous modification* for a Wiener functional is defined similarly and it holds that any two ∞–quasi continuous modications coicide everywhere except on a slim set.

It is known (cf. [22],[29]) that if $F \in \mathcal{D}_p^r$, then F possesses a (p,r)–quasi continuous modification. In particular, $F \in \mathcal{D}_{\infty-}^{\infty}$ possesses an ∞–quasi continuous modication. Combining our results so far with those obtained by Sugita [29] and Airault–Malliavin [1] we have the following :

Theorem 3.1. *Let $F : W \to R^d$ be a regular d–dimensional Wiener map. Then, for every $x \in SSP(F)$, there exists a unique Borel probability measure ν_x on W (by Borel, we mean that the domain of definition of ν_x is ν_x–completion of the topological σ–field on W) with the following properties :*

(i) *$\nu_x(A) = 0$ for every $A \subset W$ such that $C_{p,r}(A) = 0$ for some $p \in (1,\infty)$ and $r \geq [d/2]+1$.*

(ii) If *$G \in \mathcal{D}_p^r$ for some $p \in (1,\infty)$ and $r \geq [d/2]+1$, then $\tilde{G} \in L_1(\nu_x)$ and the following holds;*

$$E[G\cdot\delta_x(F)] = \int_W \tilde{G}(w)\nu_x(dx)\cdot p_F(x) \tag{3.5}$$

where $\tilde{G}$ is a (p,r)–quasi continuous modification of G.
Furthermore, ν_x is supported on $\{w; \tilde{F}(w) = x\}$, i.e.,

$$\nu_x[\{w;\tilde{F}(w) = x\}^c] = 0$$

where $\tilde{F}$ is ∞–quasi continuous modification of F.

By (3.5), it is easy to see that

$$\int_E \{\int_{\{\tilde{F}(w)=x\}} \tilde{G}(w)\nu_x(dw)\}\cdot p_F(x)dx = E[G\cdot I_E(F)],\ E \in \mathcal{B}(R^d),$$

where G and $\tilde{G}$ are as above and hence, the family $\{\nu_x\}_{x\in SSP(F)}$ defines a version of conditional probability $P(\ |F = x)$. The map $x \in SSP(F) \to \nu_x$ is

C^∞ in the sense that if $G \in \mathcal{D}_{1+}^{\infty}$ and $\tilde{G}$ is its quasi continuous modification, $x \to \int_{\{\tilde{F}(w)=x\}} \tilde{G}(w)\nu_x(dw)$ is C^∞.

Finally, we will discuss how a new approach to pinned diffusions can be obtained by applying our results to Itô functionals, i.e. Wiener functionals obtained from solutions of SDE's. Although the following can be extended to SDE's on manifolds, we restrict ourselves to SDE's on R^d for simplicity.

So let

$$
\begin{aligned}
dX_t &= \sum_{\alpha=1}^{r} V_\alpha(X_t) \circ dw^\alpha(t) + V_0(X_t)dt \\
X_0 &= x
\end{aligned}
\tag{3.6}
$$

be a SDE on R^d defined on the r–dimensional Wiener space (W,P) in the Stratonovich form where

$$V_\alpha(x) = \sum_{i=1}^{d} V_\alpha^i(x)\frac{\partial}{\partial x_i}, \quad \alpha = 0,1,\ldots,r$$

are vector fields on R^d. We assume that $V_\alpha^i(x)$ are C^∞ with bounded derivatives of all orders. Let $X(t,x,w)$ be the solution of (3.6). Then, for fixed $t \geq 0$ and $x \in R^d$, $X(t,x,\cdot) \in \mathcal{D}_{\infty-}^{\infty}(R^d)$. Furthermore, by the works initiated by Malliavin and improved much by Kusuoka and Stroock [15], we know that $X(t,x,w)$, for a fixed $t > 0$ and $x \in R^d$, satisfies (2.2) under a certain condition of Hörmander type for Lie algebra generated by vector fields V_α.

Suppose that $X(1,x,w)$ satisfies (2.2). Then the d–dimensional Wiener map $w \in W \to X(1,x,w) \in R^d$ is regular and hence, for every $y \in SSP(X(1,x,\cdot))$, the probability measure ν_y can be constructed by Th. 3.1. Let G be any Wiener functional with values in a topological space which possesses (p,r)–quasi continuous modification for some $p > 1$ and $r \geq [d/2] + 1$. *We always represent it by its quasi continuous modification with the same notation* G. Then G can be regarded as a random element on the probability space (W,ν_y), $y \in SSP(X(1,x,\cdot))$, or on the space $(\{w; X(1,x,w) = y\}, \nu_y)$. Hence every solution of SDE defined on the Wiener space (W,P) can be regarded as a stochastic process over this probability space $(\{w; X(1,x,w) = y\}, \nu_y)$ because it possesses an ∞–quasi continuous modification (Sugita [29], Example 4.2). In particular, $X(\cdot,x,w)$ can be considered as a stochastic process on this probability space which is clearly a realization of pinned diffusion process conditioned so that $X(1,x,w) = y$.

Also any property for random elements of this type which holds (p,r)–*quasi everywhere* (i.e. it holds everywhere except on a set of (p,r)–capacity 0) for some $p > 1$ and $r \geq [d/2] + 1$ is a property holding almost everywhere for measure ν_y. Therefore, it is important to sharpen a statement holding almost everywhere on (W,P) to a statement holding quasi everywhere. For example, it can be shown that the diffeomorphic property of solutions of SDE's holds not only almost everywhere but also ∞–quasi everywhere. We can reasonably expect that such a sharpening is always possible for almost important properties of solutions to SDE's. We remark that this problem of sharpening has been extensively studied for almost sure sample properties of Wiener processes (e.g. [7],[31]).

4. Asymptotic evaluation of generalized Wiener functional integrations

Many important problems in analysis, geometry and mathematical physics, such as asymptotics of eigenvalues of Laplacian, index theorems, fixed point formulas, Morse inequalities for Morse functions, Poincaré–Hopf index theorem for vector fields, etc., are essentially related to problems of estimating the traces (supertraces) of heat kernel in short time. The approaches by Wiener functional integrations to this problem consist of first representing the heat kernel by the generalized expectation of a generalized Wiener functional and then studying the asymptotics of the functional by probabilistic techniques. Very roughly, the method systematically used may be described as follows : We represent the quantity $p(\epsilon)$ (typically a heat kernel or supertrace of a heat kernel) for which we would estimate the asymptotic with respect to a parameter ϵ as $p(\epsilon) = E(\Phi(\epsilon))$ by the generalized expectation of $\Phi(\epsilon, w) \in \mathcal{D}_{1+}^{-\infty}$ depending on the parameter ϵ. We decompose $\Phi(\epsilon)$ as $\Phi(\epsilon) = \Phi_1(\epsilon) + \Phi_2(\epsilon)$ in such a way that, first for Φ_1, we can estimate its Sobolev norm and thereby show that its generalized expectation is negligible and, secondly for Φ_2, we can obtain an expansion (as explained below);

$$\Phi_2(\epsilon, w) = \varphi_0(w) + \epsilon\varphi_1(w) + \ldots + \epsilon^n\varphi_n(w) + O(\epsilon^{n+1}) \text{ in } \mathcal{D}_{1+}^{-\infty} \text{ as } \epsilon \downarrow 0$$

so that by taking the generalized expectation, we have

$$E(\Phi_2(E)) = E(\varphi_0) + \epsilon E(\varphi_1) + \ldots + \epsilon^n E(\varphi_n) + O(\epsilon^{n+1}) \text{ as } \epsilon \downarrow 0.$$

The first step is usually achieved by applying integration by parts and the theory of large deviations on Wiener space.

We now introduce relevant notions used in the second step. Let $\{\Phi(\epsilon, w)\}$ be a family $\Phi(\epsilon, \cdot) \in \mathcal{D}_{1+}^{-\infty}(E)$ of generalized E-valued Wiener functionals (E is a real separable Hilbert space as before) depending on a parameter $\epsilon \in (0,1]$. We say

$$\Phi(\epsilon, w) = O(\epsilon^k) \quad (o(\epsilon^k)) \text{ as } \epsilon \downarrow 0 \text{ in } \mathcal{D}_p^s(E)$$

if

$$\limsup_{\epsilon\downarrow 0} \|\Phi(\epsilon, w)\|_{p,s;E}/\epsilon^k < \infty \quad (\text{resp.} = 0).$$

Here we understand that $\|\Phi(\epsilon, w)\|_{p,s;E} = \infty$ if $\Phi(\epsilon, w) \notin \mathcal{D}_p^s(E)$. Based on this notion, it is natural to define the following :

(i) $\Phi(\epsilon, w) = O(\epsilon^k)$ in $\mathcal{D}_{\infty-}^{\infty}(E)$ as $\epsilon \downarrow 0$ if for every $s > 0$ and $1 < p < \infty$, $\Phi(\epsilon, w) = O(\epsilon^k)$ in $\mathcal{D}_p^s(E)$ as $\epsilon \downarrow 0$,

(ii) $\Phi(\epsilon, w) = O(\epsilon^k)$ in $\mathcal{D}_{1+}^{\infty}(E)$ as $\epsilon \downarrow 0$ if for every $s > 0$ there exists $1 < p < \infty$ such that $\Phi(\epsilon, w) = O(\epsilon^k)$ in $\mathcal{D}_p^s(E)$ as $\epsilon \downarrow 0$,

(iii) $\Phi(\epsilon,w)=O(\epsilon^k)$ in $D_{\infty-}^{-\infty}(E)$ as $\epsilon\downarrow 0$ if $s>0$ exists such that $\Phi(\epsilon,w)=O(\epsilon^k)$ in $D_p^{-s}(E)$ as $\epsilon\downarrow 0$ for all $1<p<\infty$,

(iv) $\Phi(\epsilon,w)=O(\epsilon^k)$ in $D_{1+}^{-\infty}(E)$ as $\epsilon\downarrow 0$ if $\Phi(\epsilon,w)=O(\epsilon^k)$ in $D_p^{-s}(E)$ as $\epsilon\downarrow 0$ for some $s>0$ and $1<p<\infty$.

Then it is easy to see from (1.23) and (1.25) that, listing as

$$
\begin{aligned}
F(\epsilon,w)=O(\epsilon^k)\quad \text{as}\quad \epsilon\downarrow 0\quad &\text{in}\ \ D_{\infty-}^{\infty}\ \ldots && (a)\\
&\text{in}\ \ D_{1+}^{\infty}\ \ldots && (b)\\
&\text{in}\ \ D_{\infty-}^{-\infty}\ \ldots && (c)\\
&\text{in}\ \ D_{1+}^{-\infty}\ \ldots && (d)
\end{aligned}
$$

and

$$
\begin{aligned}
\Phi(\epsilon,w)=O(\epsilon^m)\quad \text{as}\quad \epsilon\downarrow 0\quad &\text{in}\ \ D_{\infty-}^{\infty}(E)\ldots && (A)\\
&\text{in}\ \ D_{1+}^{\infty}(E)\ldots && (B)\\
&\text{in}\ \ D_{\infty-}^{-\infty}(E)\ldots && (C)\\
&\text{in}\ \ D_{1+}^{-\infty}(E)\ldots && (D),
\end{aligned}
$$

we have

$$
\begin{aligned}
F(\epsilon,w)\Phi(\epsilon,w)&=O(\epsilon^{k+m})\quad \text{as}\quad \epsilon\downarrow 0\\
&\text{in}\ \ D_{\infty-}^{\infty}(E)\ \ \text{if (a), (A) hold,}\\
&\text{in}\ \ D_{1+}^{\infty}(E)\ \ \text{if (a), (B) or (b), (A) hold,}\\
&\text{in}\ \ D_{\infty-}^{-\infty}(E)\ \ \text{if (a), (C) or (c), (A) hold,}\\
&\text{in}\ \ D_{1+}^{-\infty}(E)\ \ \text{if one of the following hold;}\\
&\text{(a), (D) or (b), (C) or (c), (B) or (d), (A).}
\end{aligned}
$$

We see by (1.26) that if $\Phi(\epsilon,w)=O(\epsilon^k)$ as $\epsilon\downarrow 0$ in $D_{1+}^{-\infty}$, then

$$E(\Phi(\epsilon,w))=O(\epsilon^k)\quad \text{as}\ \ \epsilon\downarrow 0$$

in the ordinary numerical sense. We say

$$\Phi(\epsilon,w)\sim\varphi_0+\epsilon\varphi_1+\ldots \text{in}\ D_{\infty-}^{\infty}(E)\ (D_{1+}^{\infty}(E),\ D_{\infty-}^{-\infty}(E),\ D_{1+}^{-\infty}(E))\ \text{as}\ \epsilon\downarrow 0,$$

if $\varphi_0,\varphi_1,\ldots\in D_{\infty-}^{\infty}(E)$ (resp. $D_{1+}^{\infty}(E)$, $D_{\infty-}^{-\infty}(E)$, $D_{1+}^{-\infty}(E)$) and for every $n=0,1,\ldots,$

$$
\begin{aligned}
\Phi(\epsilon,w)-[\varphi_0+\epsilon\varphi_1+\ldots\epsilon^n\varphi_n]&=O(\epsilon^{n+1})\quad \text{as}\ \ \epsilon\downarrow 0\\
\text{in}\ \ D_{\infty-}^{\infty}(E)\ \ &(\text{resp.} D_{1+}^{\infty},\ D_{\infty-}^{-\infty}(E),\ D_{1+}^{-\infty}(E)).
\end{aligned}
$$

Suppose that we are given a family of regular d-dimensional Wiener maps $F(\epsilon)=F(\epsilon,w)$ depending on $\epsilon\in(0,1]$. Assume further that it is *uniformly nondegenerate* in the sense

(4.1) $$\limsup_{\epsilon\to 0}\|(\det\sigma_{F(\epsilon)})^{-1}\|_p<\infty\quad \text{for all}\ \ 1<p<\infty$$

and admits the asymptotic expansion

$$F(\epsilon, w) \sim f_0 + \epsilon f_1 + \dots \quad \text{in } D^{\infty}_{\infty -}(R^d) \quad \text{as } \epsilon \downarrow 0 \tag{4.2}$$

where $f_i \in D^{\infty}_{\infty -}(R^d)$. Then setting $F(0,w) = f_0(w)$, $F(0,w)$ is also regular with

$$\|(\det \sigma_{F(0)})^{-1}\|_p = \lim_{\epsilon \to 0} \|(\det \sigma_{F(\epsilon)})^{-1}\|_p.$$

Theorem 4.1. ([36]) *Under the above assumptions, we have for every $T \in S'(R^d)$ that $T \circ F(\epsilon)$ has the asymptotic expansion in $D^{-\infty}_{\infty -}$ (a fortiori in $D^{-\infty}_{1+}$)*

$$T \circ F(\epsilon) \sim \varphi_0 + \epsilon\varphi_1 + \dots \quad \textit{in } D^{-\infty}_{\infty -} \quad \textit{as } \epsilon \downarrow 0 \tag{4.3}$$

and $\varphi_0, \varphi_1, \dots \in D^{-\infty}_{\infty -}$ are determined by the formal Taylor expansion

$$\begin{aligned} &T(f_0 + [\epsilon f_1 + \epsilon^2 f_2 + \dots]) \\ &= \sum_n \frac{1}{n!}(\partial^n T)(f_0) \cdot [\epsilon f_1 + \epsilon^2 f_2 + \dots]^n \end{aligned} \tag{4.4}$$

where $n = (n_1, \dots, n_d)$ is a multi-index, $n! = n_1! \cdot \dots \cdot n_d!$,

$$\begin{aligned} a^n &= a_1^{n_1} \cdot \dots \cdot a_d^{n_d} \quad \textit{for } a = (a_1, \dots, a_d) \in R^d \quad \textit{and} \\ \partial^n &= (\partial/\partial x_1)^{n_1} \dots (\partial/\partial x_d)^{n_d}. \end{aligned}$$

In particular, denoting $\partial^i = \partial/\partial x^i$,

$$\varphi_0 = T(f_0), \quad \varphi_1 = \sum_{i=1}^d f_1^i \cdot (\partial^i T)(f_0),$$

$$\varphi_2 = \sum_{i=1}^d f_2^i \cdot (\partial^i T)(f_0) + \frac{1}{2!} \sum_{i,j=1}^d f_1^i \cdot f_1^j \cdot (\partial^i \partial^j T)(f_0),$$

$$\begin{aligned} \varphi_3 = \sum_{i=3}^d f_3^i \cdot (\partial^i T)(f_0) &+ \frac{2}{2!} \sum_{i,j=1}^d f_1^i \cdot f_2^j \cdot (\partial^i \partial^j T)(f_0) \\ &+ \frac{1}{3!} \sum_{i,j,k=1}^d f_1^i \cdot f_1^j \cdot f_1^k \cdot (\partial^i \partial^j \partial^k T)(f_0), \end{aligned}$$

and so on.

Another similar result under somewhat different assumption is as follows :

Theorem 4.2. *Suppose that $F(\epsilon) = F(\epsilon, w)$ be a family of regular d-dimensional Wiener maps depending on a parameter $\epsilon \in [0,1]$ which satisfies (4.1) and that the map $\epsilon \in [0,1] \to F(\epsilon) \in D^{\infty}_{\infty-}(R^d)$ is C^1 (as a Frechet space valued function). Then, for every $T \in S'(R^d)$, $T \circ F(\epsilon)$ is defined by Th. 2.1 and, for every $k = 0, 1, \ldots$, we have*

$$T \circ F(\epsilon) = \sum_{n, |n| \leq k} \frac{1}{n!} (\partial^n T)(F(0)) \cdot (F(\epsilon) - F(0))^n + O(\epsilon^{k+1}) \tag{4.5}$$

for $D^{-\infty}_{\infty-}$ as $\epsilon \downarrow 0$.

Here $|n| = n_1 + \ldots n_d$ for $n = (n_1, \ldots, n_d)$.

Proof is easily provided by applying the following formula successively :

$$T \circ F(\epsilon) - T \circ F(0) = \sum_{i=1}^{d} \int_0^{\epsilon} (\partial^i T) \circ F(u) \cdot \frac{dF^i}{du}(u) du.$$

As a typical application, we consider a SDE on R^d defined over the Wiener space (W, P) with parameter $\epsilon \in [0,1]$:

$$dX_t = \epsilon \sum_{\alpha=1}^{r} V_\alpha(X_t) \circ dw^\alpha(t) + \epsilon^2 V_0(X_t) dt$$
$$X_0 = x$$

where the vector fields satisfy the same conditions as in §3. Let $X^\epsilon(t, x, w)$ be the solution. It follows from the scaling property of the Wiener measure that $\{X^\epsilon(t, x, w)\}$ and $\{X^1(\epsilon^2 t, x, w)\}$ have the same law. It is easy to obtain the asymptotic expansion in $D^{\infty}_{\infty-}(R^d)$ of $F(\epsilon, w) = X^\epsilon(1, x, w)$ as $\epsilon \downarrow 0$ (cf. [36],[33]);

$$F(\epsilon, w) \sim f_0(w) + \epsilon f_1(w) + \ldots$$

where $f_i \in D^{\infty}_{\infty-}(R^d)$ can be explicitly computed as a finite sum of multiple Wiener–Itô integrals. In particular, $f_0(w) = x$,

$$F_1(w) = \sum_{\alpha=1}^{r} V_\alpha(x) \cdot w^\alpha(1) \text{ and } f_2(w) = V_0(x) + \sum_{\alpha,\beta=1}^{r} \sum_{i=1}^{d} \partial^i V_\alpha(x) \cdot V^i_\beta(x)$$
$$\times \int_0^1 w^\beta(t) \circ dw^\alpha(t).$$

If $\det A(x) > 0$, where $A^{ij}(x) = \sum_{\alpha=1}^{r} V^i_\alpha(x) V^j_\alpha(x)$, then $F(\epsilon)$ is regular for each $\epsilon \in (0,1]$ and $G(\epsilon, w) = \frac{F(\epsilon,w)-x}{\epsilon}$ satisfies (4.1). Hence by Th.4.1 or 4.2 we have

$$\delta_x(F(\epsilon, w)) = \epsilon^{-d} \delta_0(G(\epsilon, w)) \sim \epsilon^{-d}(\varphi_0 + \epsilon\varphi_1 + \ldots) \text{ in } D^{-\infty}_{\infty-}$$

as $\epsilon \downarrow 0$ and can compute $\varphi_i \in D_{\infty-}^{-\infty}$ explicitly. In particular,

$$\varphi_0(w) = \delta_0(f_1(w)) = (\det A(x))^{-1/2}\delta_0(w(1)).$$

These results can be applied to obtain short time asymptotic expansion of heat kernels on the diagonal and thereby obtain probabilistic proofs of index theorems, Morse inequalities, etc. (cf. e.g. [36],[33],[11],[20],[26],[32]). We should remark that the original work is due to Bismut ([4],[5],[6]) and his method is refined and expanded by Kusuoka [13].

In the case of heat kernels off the diagonal, we need to consider the Wiener functional $F(\epsilon, w) = X^\epsilon(1, w, w + h/\epsilon)$ for a suitably chosen $h \in H$, cf. [13],[36], [3], [19].

Also, we can study by the same method the case of $\det A(x) = 0$ under a certain hypoellipticity condition on vector fields, cf. [18],[3],[19],[30],[34]. The case of heat kernels on a manifold with boundary can be discussed by a similar method by modifying the Malliavin calculus in a suitable way cf. [9],[27],[37].

References

[1] H. Airault et P. Malliavin, Intégration géométrique sur l'espace de Wiener, *Bull. Sci. Math.*, 2e série, **112** (1988), 3-52.

[2] R. Azencott, Densité des diffusions en temps petit : développements asymptotiques, in *Séminaire de Prob. XVIII*, 1982/1983, (ed. J. Azéma et M. Yor), **LNM.1059**, Springer (1984), 402-498.

[3] G. Ben Arous, Noyau de la chaleur hypoelliptique et géométrie sous–riemanniene, in *Stochastic Analysis, Proc. Paris* 1987, (ed. M. Métivier and S. Watanabe), **LNM.1322**, Springer (1988), 1-16.

[4] J-M. Bismut, Large deviations and the Malliavin calculus, *Progress in Math.*, **Vol.45**, Birkhauser, 1984.

[5] J-M. Bismut, The Atiyah–Singer theorems : a probabilistic approach, I, the index theorem, II, the Lefschetz fixed point formulas, *J. Funct. Anal.*, **57** (1984), 56-99 and 329-348.

[6] J-N. Bismut, The Witten complex and degenerate Morse inequalities, *J. Diff. Geom.*, **23** (1986), 207-240.

[7] M. Fukushima, Basic properties of Brownian motion and capacity on the Wiener space, *J. Math. Soc. Japan*, **36** (1984), 161-176.

[8] P. Hsu, Brownian motion and Riemannian geometry, in *Geometry of Random motion, Proc. Summer Research Conf.* 1987, (ed. R. Durret and M. A. Pinsky), Contemporary Math. **Vol.73**, AMS (1988), 95-104.

[9] N. Ikeda and S. Kusuoka, Short time asymptotics for fundamental solutions of diffusion equations, in *Stochastic Analysis, Proc. Paris* 1987, (ed. M. Métivier and S. Watanabe), **LNM.1322**, Springer (1988), 37-49.

[10] N. Ikeda and S. Watanabe, *Stochastic differential equations and diffusion processes*, Kodansha/North-Holland, 1981, 2nd edition, 1989.

[11] N. Ikeda and S. Watanabe, Malliavin calculus of Wiener functionals and its applications, in *From local times to global geometry, control and physics, Warwick Symp. on SDE and Appl.* 1984/85, (ed. K. D. Elworthy), Pitman Research Notes in Math. Ser. **150**, Longman (1986), 132-178.

[12] K. Itô, Malliavin calculus on a Segal space, in *Stochastic Analysis, Proc. Paris* 1987, (ed. M. Métivier and S. Watanabe), **LNM.1322**, Springer (1988), 50-72.

[13] S. Kusuoka, The generalized Malliavin calculus based on Brownian sheet and Bismut's expansion for large deviation, in *Stochastic Processes-Mathematics and Physics, Proc. 1st Bibos Symp.*, (ed. S. Albeverio, Ph. Blanchard and L. Streit), **LNM.1158**, Springer (1986), 141-157.

[14] S. Kusuoka and D. W. Stoock, Applications of the Malliavin calculus, Part I, in *Stochastic Analysis, Proc. Taniguchi Symp. Katata and Kyoto* 1982, (ed. K. Itô), Kinokuniya/North Holland (1984), 271-306.

[15] S. Kusuoka and D. W. Stroock, Applications of the Malliavin calculus, Part II, *J. Fac. Sci. Univ. Tokyo, Sect. 1A. Math.*, **32** (1985), 1-76.

[16] S. Kusuoka and D. W. Stroock, Applications of the Malliavin calculus, Part III, *J. Fac. Sci. Univ. Tokyo, Sect. 1A, Math.*, **34** (1987), 391-442.

[17] H. H. Kuo, Donsker's delta function as a generalized Brownian functional and its application, in *Theory and Applications of Random Fields, Proc. IFIP-WG 7/1 Working Conf.* (ed. G. Kallianpur) **LNCI.49**, Springer (1983), 167-178.

[18] R. Léandre, Développement asymptotique de la densité d'une diffusion dégénérée, to appear in *J. Funct. Anal.*

[19] R. Léandre, Applications quantititatives et géométriques du calcul de Malliavin, in *Stochastic Analysis, Proc. Paris* 1987 (ed. M. Métivier and S. Watanabe), **LNM.1322**, Springer (1988), 109-133.

[20] R. Léandre, Sur le théorème d'Atiyah-Singer, *Probab. Th. Rel. Fields*, **86** (1988), 119-137.

[21] P. Malliavin, Stochastic calculus of variation and hypoelliptic operators, in *Stochastic Differential Equations, Proc. Int. Symp.* SDE Kyoto 1976, (ed. K. Itô), Kinokuniya (1987), 155-263.

[22] P. Malliavin, Implicit functions in finite corank on the Wiener space, in *Stochastic Analysis, Proc. Taniguchi Symp. Katata and Kyoto* 1982, (ed. K. Itô), Kinokuniya (1984), 369-386.

[23] P. A. Meyer et J. A. Yan, A propos des distributions sur l'espace de Wiener, in *Séminaire de Prob. XXI*, (ed. J. Azéma, p. A. Meyer et M. Yor), **LNM.1247**, Springer (1987), 8-26.

[24] P. A. Meyer et J. A. Yan, Distirbutions sur l'espace de Wiener (suite) d'après I. Kubo et Y. Yokoi, in *Séminaire de Prob. XXIII*, (ed. J. Azéma, P. A. Meyer, M. Yor), **LNM.1372**, Springer (1989), 382-392.

[25] S. A. Molchanov, Diffusion processes and Riemannian geometry, *Russ. Math. Surveys*, **30** (1975), 1-63, From Uspekhi Mat. Nauk, **30** (1975), 3-59.

[26] I. Shigekawa and N. Ueki, A stochastic approach to the Riemann–Roch theorem, *Osaka J. Math.*, **25** (1985), 717-725.

[27] I. Shigekawa, N. Ueki and S. Watanabe, A probabilistic proof of the Gauss–Bonnet–Chern theorem for manifolds with boundary, to appear in *Osaka J. Math.*

[28] H. Sugita, Sobolev spaces of Wiener functionals and Malliavin calculus, *J. Math. Kyoto Univ.*, **25** (1985), 31-48.

[29] H. Sugita, Positive generalized Wiener functions and potential theory over abstract Wiener spaces, *Osaka J. Math.*, **25** (1988), 665-696.

[30] S. Takanobu, Diagonal short time asymptotics of heat kernels for certain degenerate second order differential operators of Hörmander type, *Publ. RIMS, Kyoto Univ.*, **24** (1988), 167-203.

[31] M. Takeda, (r,p)–capacity on the Wiener space and properties of Brownian motion, *Z. Wahrsch. verw. Geb.*, **68** (1984), 149-162.

[32] N. Ueki, A stochastic approach to the Poincaré–Hopf theorem, *Probab. Th. Rel. Fields*, **82** (1989), 271-293.

[33] H. Uemura, On a short time expansion of the fundamental solution of heat equations by the method of Wiener functionals, *J. Math. Kyoto Univ.*, **27** (1987), 417-431.

[34] H. Uemura and S. Watanabe, Diffusion processes and heat kernels on certain nilpotent groups, in *Stochastic Analysis, Proc. Paris* 1987, (ed. M. Métivier and S. Watanabe), **LNM.1322** Springer (1988), 173-197.

[35] S. Watanabe, *Lectures on stochastic differential equations and Malliavin calculus*, Tata Institute of Fundamental Research/Springer, 1984.

[36] S. Watanabe, Analysis of Wiener functionals (Malliavin calculus) and its applications to heat kernels, *Ann. Probab.*, **15** (1987), 1-39.

[37] S. Watanabe, A modified Malliavin calculus and its applications to heat kernels with boundary conditions, in preparation.

Appendices

List of contributors

Roger GAY, Warrnambool Institute of Advanced Education, P O Box 423, Warrnambool, Vic. 3280, Australia

*C. C. HEYDE**, School of Mathematical Sciences, The Australian National University, GPO Box 4, Canberra ACT 2601, Australia

*Takeyuki HIDA**, Department of Mathematics, Faculty of Science, Nagoya University, Nagoya 464, Japan

*Kiyosi ITÔ**, Research Institute of Mathematical Sciences, Kyoto University, Kyoto 606, Japan

*Harry KESTEN**, Department of Mathematics, Cornell University, Ithaca, NY 14853-7901, USA

*Shinichi KOTANI**, Department of Mathematics, University of Tokyo, Hongo, Tokyo 113, Japan

*Mark A. PINSKY**, Department of Mathematics, Northwestern University, Evanston, Illinois 60208, USA

Jürgen POTTHOFF, Department of Mathematics, Louisiana State University, Baton Rouge, Louisiana 70803, USA

*David O. SIEGMUND**, Department of Statistics, Stanford University, Stanford, California 94305, USA

*Charles STEIN**, Department of Statistics, Stanford University, Stanford, California 94305, USA

*Daniel W. STROOCK**, Department of Mathematics, Massachusetts Institute of Technology, Cambridge, Massachusetts 02139, USA

*Shinzo WATANABE**, Department of Mathematics, Faculty of Science, Kyoto University, Kyoto 606, Japan

Volker WIHSTUTZ, Department of Mathematics, University of North Carolina at Charlotte, Charlotte, North Carolina 28223, USA

*Jia-an YAN**, Institute of Applied Mathematics, Academia Sinica, Beijing, China

---

* Invited Speakers

List of participants

S.R. ADKE, Department of Statistics, University of Poona, Pune

W.J. ANDERSON, Department of Mathematics & Statistics, McGill University, Montreal

J. AZÉMA, Laboratoire de Probabilités, Université Paris VI, Paris

Paolo BALDI, Dipartimento Di Mathematica, Citta' Universitaria, Catania

A. D. BARBOUR, Institut für Angewandte Mathematik der Universität Zürich, Zürich

Thomas BIER*

Timothy BROWN, Department of Mathematics, University of Western Australia, Nedlands

CHAN Kai Meng*

N.N. CHAN, Department of Statistics, The Chinese University of Hong Kong, Shatin

CHAN Onn, Baltimore

CHAN Yiu Man*

CHANG Boon Chong, Department of Mathematics, National University of Singapore, Singapore

Kun Soo CHANG, Department of Mathematics, Yonsei University, Seoul

CHAO Min-Te, Institute of Statistical Science, Academia Sinica, Taipei

Louis H.Y. CHEN*

CHENG Kai Nah*

---

*Department of Mathematics, National University of Singapore, Singapore

CHENG Kuang-Fu, Graduate Institute of Statistics, National Central University, Chung-li

CHEW Tuan Seng*

CHIANG Tzuu-Shuh, Institute of Mathematics, Academia Sinica, Taipei

CHIEW Mui Kui, School of Arts & Science, Tunku Abdul Rahman College, Kuala Lumpur

CHOI Kwok Pui*

CHONG Chi Tat*

CHOU Ching-Sung, Institute of Mathematics, National Central University, Chung-li

CHOW Yoong Ming*

CHOW Yunshyong, Institute of Mathematics, Academia Sinica, Taipei

CHUA Kok Seng*

CHUA Seng Kiat, Department of Mathematics, Brown University, Providence

Dong Myung CHUNG, Department of Mathematics, Sogang University, Seoul

DAI Yonglong, Department of Mathematics, Zhongshan University, Guangdong

Robert C. DALANG, Department of Statistics, University of California, Berkeley

Persi DIACONIS, Department of Mathematics, Harvard University, Cambridge

E.B. DYNKIN, Department of Mathematics, Cornell University, Ithaca

EE Soo Mei, Department of Mathematics, University of Malaya, Kuala Lumpur

GAN Fah Fatt*

GOH Beng Huay*

GONG Guanglu, Department of Applied Mathematics, Tsinghua University, Beijing

R.C. GUPTA*

Adolfo M. DE GUZMAN, Quezon City

---

*Department of Mathematics, National University of Singapore, Singapore

HANG Kim Hoo, Singapore

HE Shengwu, Department of Mathematical Statistics, East China Normal University, Shanghai

C.C. HEYDE, School of Mathematical Sciences, The Australian National University, Canberra

Takeyuki HIDA, Department of Mathematics, Nagoya University, Nagoya

Lars HOLST, Department of Mathematics, Royal Institute of Technology, Stockholm

HONG Geck Chan*

HOU Zhen-ting, Changsha Railway Institute, Hunan

HU Kaiyuan*

HUANG Zhirui, Department of Mathematics, Zhongshan University, Guangdong

HWANG Chii-Ruey, Institute of Mathematics, Academia Sinica, Taipei

Yasushi ISHIKAWA, Department of Information Science, Hirosaki University, Hirosaki

Hiroshi ISHITANI, Department of Mathematics, Mie University, Mie

Kiyosi ITÔ, Research Institute of Mathematical Sciences, Kyoto University, Kyoto

Jong Woo JEON, Department of Computer Science & Statistics, Seoul National University, Seoul

Mamoru KANDA, Department of Mathematics, University of Tsukuba, Ibaraki

Harry KESTEN, Department of Mathematics, Cornell University, Ithaca

KHATIJAH Syed Ahmad, Pusat Pengajian Sains Matematik & Sains Komputer, Universiti Sains Malaysia, Pulau Pinang

KOH Khee Meng*

Shinichi KOTANI, Department of Mathematics, University of Tokyo, Tokyo

LAM Yeh, Department of Statistics, The Chinese University of Hong Kong, Shatin

Andy H. LEE , School of Mathematics, Northern Territory University, Casuarina

---

*Department of Mathematics, National University of Singapore, Singapore

LEE Seng Hwang*

LEE Seng Luan*

LEE Soo Teck*

LEONG Yu Kiang*

LEUNG Ka Hin*

LEUNG Pui Fai, Fred*

LIM Keng Suan*

LIM Siok Wan, School of Arts & Science, Tunku Abdul Rahman College, Kuala Lumpur

LIN Ta-Feng, Department of Mathematics, Soochow University, Taipei

LING Kiong Doong*

Susanti LINUWIH, Surabaya

LIU Wen, Department of Mathematics, Hebei Institute of Technology, Tianjin

LIU Xiufang, Department of Mathematics, Beijing Normal University, Beijing

LOH Wing Wah*

Jiann-Hua LOU*

MA Siu Lun*

MAK Kok Sun*

Brett T. MCINNES*

Minoru MOTOO, Tokyo

Joseph K. MOURAD*

M. Gopalan NAIR, Department of Algebra, Combinatorics & Analysis, Auburn University, Auburn

Shizu NAKANISHI, Osaka

T. R. NANDA*

*Department of Mathematics, National University of Singapore, Singapore

Kritsana NEAMMANEE, Department of Mathematics, Chulalongkorn University, Bangkok

David Kwok Wai NG, Department of Mathematics, Eastern Kentucky University, Kentucky

NG Tze Beng*

Masami OKADA, Department of Mathematics, Tôhoku University, Sendai

Krzysztof OSTASZEWSKI, Department of Mathematics, University of Louisville, Louisville

PANG Yu Hin, Peter*

PHANG Lay Ping, Rosalind*

Mark A. PINSKY, Department of Mathematics, Northwestern University, Evanston

POH Kheng Siong, Roger*

QUEK Tong Seng*

Malcolm QUINE, Department of Mathematical Statistics, University of Sydney, Sydney

Rm. SEKKAPPAN, Department of Mathematics, University of Papua New Guinea, Papua New Guinea

SHEE Sze Chin*

SHEN Anqi, Shenzhen Education College, Shenzhen

Narn-Rueih SHIEH, Department of Mathematics, National Taiwan University, Taipei

David O. SIEGMUND, Department of Statistics, Stanford University, Stanford

Karl SIGMAN, Department of Industrial Engineering & Operations Research & The Ctr for Telecomms Research, Columbia University, New York

Saisuda SOMCHIT, Bangkok

SOON Yu Tiong*

Charles STEIN, Department of Statistics, Stanford University, Stanford

---

*Department of Mathematics, National University of Singapore, Singapore

Daniel W. STROOCK, Department of Mathematics, Massachusetts Institute of Technology, Cambridge

Shigeo TAKENAKA, Department of Mathematics, Nagoya University, Nagoya

TANG Wai Shing*

TAY Kian Boon*

TAY Tiong Seng*

Premjai TRISARANUWATANA, Bangkok

TUNG Chi Tung, Peter*

Jose S. VICTORIA, College of Business Administration, University of the Philippines, Quezon City

Angela WADE, Department of Community Medicine, London

WANG Ann Lee, Department of Mathematics, University of Malaya, Kuala Lumpur

Shinzo WATANABE, Department of Mathematics, Kyoto University, Kyoto

WONG Char Ngan*

WONG Woon Kwong*

XIE Shengrong, Department of Mathematics, Southwest-China Normal University, Chongqing

XU Jia Gu, Department of Mathematics, Fudan University, Shanghai

YAN Jia-an, Institute of Applied Mathematics, Academia Sinica, Beijing

YANG Zhen Ming, Department of Mathematics, Nankai University, Tianjin

YAP Kim Yew*

YEO Kwee Poo, Statistics Centre, Cornell University, Ithaca

ZALILA bte Ali, Centre for Mathematical & Computer Sciences, Universiti Sains Malaysia, Pulau Pinang

---

*Department of Mathematics, National University of Singapore, Singapore

Bei Fragen zur Produktsicherheit wenden Sie sich bitte an:
If you have any questions regarding product safety,
please contact:

Walter de Gruyter GmbH
Genthiner Straße 13
10785 Berlin
productsafety@degruyterbrill.com